Praise for *The Power of Babel*

"Startling, provocative, and remarkably entertaining. McWhorter's prose crackles [and] his pop-culture references pop. His enthusiasm for his topic is infectious. . . . McWhorter displays impressive literary footwork."
—*San Diego Union-Tribune*

"*The Power of Babel* is sprawling, gossipy, and fascinating. McWhorter has a skill at answering questions you have never thought to ask: When Ginger Rogers said that Fred Astaire 'made love to me' in a 1935 movie, what precisely did she mean? Why is Charlie Brown bald? Do Germans fully appreciate the translated version of the American TV sitcom *Married . . . with Children*? Why does Mickey Mouse wear white gloves? Surprisingly, the answers bear in interesting ways on the nature of language and linguistic diversity."
—*American Scientist* magazine

"John McWhorter's *The Power of Babel: A Natural History of Language* is an essay in origins, and is as theoretical as Hawking and Gorst in trying to see into the deep past. McWhorter is a clear and witty writer."
—*Harper's* magazine

"McWhorter explains clearly how and why sounds change, how word meanings change . . . how grammar changes and how they all bifurcate, mix, multiply, grow branches, get elaborated, are dissolved and reconstituted. McWhorter writes lucidly; it's evident that he's a teacher."
—*San Francisco Chronicle*

"McWhorter's use of analogies, anecdotes, and popular culture keeps the discussion lively. A worthy contribution to our understanding of the defining feature of human life."
—*Booklist*

"McWhorter offers fascinating detail. [He] has written a valuable book."
—*Sun-Sentinel* (Ft. Lauderdale, FL)

"Teeming with interesting observations. Those fascinated with languages will find [McWhorter's] book a treat—engaging and highly informative."
—*Columbus Dispatch*

"Far from being a dry, fusty, academic text, McWhorter's work is a passionate, engaging, and refreshing look at the rich history of language."
–Associated Press

"[A] far-ranging, lively excursion through the nature and history of language. There may be 6,000 tongues in the world, but only one word for this . . . fascinating."
–*Washington Post*

"Anyone even remotely interested in linguistics, history, current events . . . will find *The Power of Babel* compelling. . . . A more learned or original guide would be hard to find for this heady trip that spans most of our past while it points toward the future."
–*Roanoke Times & World News*

"*The Power of Babel* is a witty, stimulating exploration of the essentials of linguistic change, enlivened by McWhorter's gift for unexpected analogies."
–*Chicago Tribune*

"With passionate eloquence, McWhorter makes readers glimpse the wonder of languages. The pleasure of *The Power of Babel* lies equally in his argument, the details he uses to illustrate it, and the wit and friendly energy of his writing. . . . Particularly fascinating."
–New York *Newsday*

"McWhorter's arguments are sharply reasoned, refreshingly honest, and thoroughly original, and befitting a book on language, they are lucidly and elegantly expressed."
–Steven Pinker, author of *The Language Instinct*

"With a brisk, witty style that reveals a comprehensive knowledge of music and popular culture, McWhorter rarely lets his tour wander in the tangled wood of academic jargon and arcane illustration. An entertaining, instructive Henry Higgins of a volume; it'll transform readers into enraptured Eliza Doolittles."
–*Kirkus Reviews*

"McWhorter's scholarship lays enough groundwork to keep future linguists busy for centuries. For the lay reader, that scholarship will make traveling . . . all the more wondrous." –BookPage.com

"McWhorter offers a sweeping survey of people's movements around the globe and the evolution of languages. Entertaining. . . . Accessible." –*Science News* (Washington D.C.)

Michael Sexton

About the Author

JOHN MCWHORTER is associate professor of linguistics at the University of California at Berkeley. He is author of *Word on the Street: Debunking the Myth of a "Pure" Standard English* and the bestselling *Losing the Race: Self-Sabotage in Black America*. He lives in Oakland, California.

The Power of Babel

Also by John McWhorter

The Missing Spanish Creoles: Recovering the Birth of Plantation Contact Languages

Spreading the Word: Language and Dialect in America

Towards a New Model of Creole Genesis

Word on the Street: Debunking the Myth of a "Pure" Standard English

Losing the Race: Self-Sabotage in Black America

The Power of Babel

A Natural History of Language

John H. McWhorter

Perennial
An Imprint of HarperCollins*Publishers*

A hardcover edition of this book was published in 2001 by Times Books, a division of Henry Holt and Company, LLC. It is here reprinted by arrangement with Henry Holt and Company, LLC.

HarperCollins books may be purchased for educational, business, or sales promotional use. For information please write: Special Markets Department, HarperCollins Publishers Inc., 10 East 53rd Street, New York, NY 10022.

First Perennial edition published 2003.

Designed by Diana Blume

Library of Congress Cataloging-in-Publication Data

McWhorter, John H.
The power of Babel : a natural history of language / John H. McWhorter.
p. cm.
Includes bibliographical references and index.
ISBN 0-06-052085-X
1. Historical linguistics. 2. Language and languages–Origin. I. Title.

P140.M34 2003
417'.7–dc21

2002034592

08 09 10 ❖/RRD 10

Contents

The Power of Babel

Introduction

I fell in love for the first time at four years old. Her name was Shirley and we were both in a piano class. She wore burgundy overalls, which I for some reason found immensely charming (I seriously doubt if she actually wore these overalls every day as if she were a *Peanuts* character, but that's how she is preserved in my memory); she had laughing brown eyes and a high-spirited yet intelligent demeanor. I was intoxicated and, as it happened, we got along quite well.

I'm not sure of exactly what kind of trajectory I imagined us to be on, but whatever it was, it was upward–until one day after a lesson when we went outside to join our respective parents. Watching her joyously greet her family, I was shocked to hear that as soon as she started talking to them, I couldn't understand what they were saying! This was the first time in my life that I had ever known that there were languages other than English, and it remains the profoundest shock I have ever encountered in my entire life. They were clearly communicating, just as I was with my mother, but *I couldn't understand what they were saying!*

For me, this was not only shocking but heartbreaking, because I felt that Shirley's newly revealed ability cut her off from me, that she had gone somewhere I couldn't go. "What are they doing, Mom?" I asked frantically. "They're speaking another language, Jughead!" she answered ("Jughead" was a pet name). "What do you mean? Where did they learn how to do that?" I persisted. Mom went over and asked Shirley's relatives politely, "Excuse me, what language are you speaking?" "We are speaking Hebrew," intoned one of them. Mom came back to me and said, "They're speaking

Hebrew." "But why don't we speak Hebrew, Mom?" She answered, "Because we're not Jewish. Can we go home now?" And so we did.

But on the way home in the car, I was so frustrated that I cried like the child that I was–partly because I felt that this revelation had lost me the girl of my dreams and partly because I was absolutely dazzled by the idea that there were ways of speaking that I could not understand, that there were *other* ways to talk, that a person could be able to talk in *two* ways, and that I had been denied that ability by not being, well, Jewish (that's as far as I could understand it at this point).

As it happened, a Hebrew school met in the late afternoons in the building where I went to school, and I became so obsessed with my language deficit that I left a note on the blackboard for the rabbi (as directed by my teacher) asking him how I, too, could learn Hebrew. He left me a flyer with the Hebrew alphabet on it (actually, for some reason, the alphabet as used for Yiddish), which was a princely thing to do considering that he could have just ignored my missive entirely. With this flyer and a cute Hebrew-language children's picture dictionary that Mom dug up somewhere for me, I learned to sound out Hebrew (I was that kind of a kid), and that was enough for me then; it didn't occur to me that I didn't really know what I was reading or that I couldn't actually put together sentences of my own.

This was the beginning of a lifelong obsession with foreign languages. The next step was realizing that Hebrew wasn't the only language not spoken in our house. In the back of a dictionary we had, there was an appendix with a good five thousand or so words translated into French, Spanish, Italian, German, Swedish, and (for some reason) Yiddish. I decided that it was imperative that "Twinkle, Twinkle, Little Star" be translated into all six but thought that all you had to do was plug in the words into their English slots. I knew nothing of conjugation or agreement or that grammars differ from language to language. I still have that first "book" that I wrote, with deathless poetry such as the Spanish rendition: *Centellear, Centellear, Pequeño estrella, Cómo yo preguntarse, Qué tu ser.* . . . As I got older I began to teach myself languages in earnest as a hobby (the key to getting decent at it is to talk to yourself in the shower) and was eventually fortunate enough to be able to support myself on my passion by becoming a linguist. To this day, I have the flyer with the Hebrew (actually Yiddish) alphabet (the "Alef-Baze") framed and hanging over my desk, as a symbol of what sparked my combination career and avocation. Although I have long since realized that our family was hardly unique in not using Hebrew in the home and that learning a language entails more than mastering a collection of undigested words, after all these years the true roots of my fascination with language remain the same as on that day on a Philadelphia sidewalk when I lost my innocence: that anything I write or say in this language can be said in about six thousand other ways, with completely different words and with grammars so different that they can almost strain the credulity of the outsider.

Yet all of these languages are spoken by members of the same subspecies, *Homo sapiens sapiens,* to accomplish the same tasks of communication of information, expression of emotion and attitude, commanding and requesting, social libation, calibration of power relations, and poetic expression. This sort of variation within the bounds of a template is analogous to hearing the different uses to which Art Tatum, Coleman Hawkins, or Lionel Hampton could put the chord sequence and basic melody of "Body and Soul" or to seeing the many faces of the little Stephen Foster song "Shortnin' Bread," depending on whether it is sung by a classically trained mezzo from the original sheet music, warbled by Ethel Mertz at a small-town concert on *I Love Lucy,* or swung by the orchestra accompanying a Bugs Bunny cartoon. Ask someone who speaks a language other than English natively how to say *I sank into the mud up to my ankles,* and figure out what the words actually mean. The variety among the words themselves is wonder enough, but the multitude of sentence patterns in which human beings can express that homely concept are Art Tatum, Vivian Vance, and beyond: *Am intrat in noroi pǎna la glezne* (I have entered in mud up to ankles [Romanian]); *Ich bin bis zu meinen Knöcheln im Schlamm versunken* (I am until the ankles in the mud sunk [German]); *Ja provalilsja v grjaz' po ščikolotku* (I sank-self in mud at ankle [Russian]); *Doro no naka ni askikubi made tsukatte shimatta* (mud of within at ankle until soaked put-away [Japanese]); *Bikwaakoganaaning ingii-apiichi-gagwaanagwajiishkiwese* (knob-bone-at I-extending to-"mudmoved" [Ojibwe, or "Chippewa"]), and so on.

Properly speaking, though, what interests me is not this variety alone, but the variety as seen within a certain context: the fact that all of this variety is the product of evolution from a single original source. It's not an accident that this aspect of language has held my obsessive attention for so long, as I have always had a bizarre native fascination with the evolutionary aspect of everything. If losing Shirley to Hebrew remains the starkest moment of revelation in my life thus far, the second starkest was when I was about eight years old and caught an episode of that magnificent antique *The Honeymooners* on television. This was my first perception of television shows as falling upon a chronological timeline, as well as my first conscious exposure to the fact that details of American social mores, customs, and fashions change in the course of a century, and I found myself transfixed by the difference between this "old" world and the world of the 1970s (at that age, that twenty-year gap in time felt the way a fifty-year one does now). This time it was my father who had to bear my frantic questions: "Why does it look so scratchy and blurry?" "Why is it in black and white?" "Why are the women dressed that way?" "Why was that joke funny?" "Why is nobody black?" "If he loves her, why does he keep yelling at her like that?" "Why did she kiss him at the end if he's so mean?" "Why don't they ever leave that room?" "Why is the music so ugly?" (My father's answers to these questions were perfect and truly worth another book in themselves, but I digress.)

What grabbed me about *The Honeymooners* was the basic question: How did you get from there to here? How did you get from the black-and-white, claustrophobic, misogynstic vaudeville of that show to one quite current at the time, the sociologically sophisticated world of *Maude,* where the women were in charge, the script was witty, and action alternated between three rooms and was in color to boot? What were the intermediate steps? (*The Dick Van Dyke Show* was one, as I would learn later.) In the same way, the variety among the world's languages is all the more marvelous in that it is the product of a process of six thousand imperceptibly gradual transformations of what began as a single language. To wit, there was a brief shining moment in human history when I would not have lost Shirley to another language, for the simple reason that only one language existed.

Of course, in this hypothetical period during which the course of true love would have run more smoothly for Shirley and me, the two of us would have been naked, hairy, of the same race, outdoors, and in Africa, and we would be much more likely to have met digging for grubs or running from something than taking piano lessons. Deducing from a combination of data from archaeology, paleontology, molecular biology, and anatomical reconstruction, we can be almost certain that the first human beings to speak language as we know it today lived in East Africa about 150,000 years ago.

What do I mean by "language as we know it"? To understand this requires awareness of two things. First, human language differs sharply in a qualitative sense from the various levels of communicative ability, marvelous in themselves, possessed by some animals. Bees can tell other bees where honey is located by a butt-waggling dance. Chimpanzees and other apes can be trained to use a rudimentary kind of sign language. Parrots have been trained to match words to concepts. Some animals have specific cries warning their comrades against predators. We have all seen how dogs can learn to recognize a dozen or so words (you had to make sure to always spell the word *walk* in the presence of one dog I knew, because otherwise even saying "I think she wanted him to take a walk on the wild side" would lead him to spend the next two minutes jumping in ecstatic frustration waiting to be taken outside).

However, human language is unique in its ability to communicate or convey an open-ended volume of concepts: we are not limited to talking about exactly where honey is, to warning each other that something is coming to try to eat us, or to matching vocalizations to fifty-odd basic concepts pertaining to our immediate surroundings and usually focusing on bananas and desire. Neither bees, chimps, parrots nor dogs could produce or perceive a sentence such as "Did you know that there are squid fifty feet and longer in the deep sea? They have only been seen as corpses washed up on beaches." Because animals can only communicate about either things in the immediate environment or a small set of things genetically programmed ("The honey is over there," "A leopard is coming," "Banana!"), they could not tell each other about giant squid even if they had seen one, nor could they "talk" about corpses even if they had seen plenty. Then there is the specificity for which human language is designed: no animal could specify that the squid have been seen in the past, rather than being seen right now, nor could they communicate the concept of "knowing" in "Did you know . . . ?"

Not only are no animals remotely capable of communication on this level (and, if you think about it, even those sentences about giant squid are not exactly Proust) but none even approximate it: there are no animals that could even pull off "Once I met a huge animal" or the concept of "washed up on" or even the concept of "once" in the sense of "one instance in the past." There is a vast gulf in complexity, subtlety, and flexibility between human beings and other animals in regard to language ability, and that gulf is a large part of why humans have been such a successful species of such disproportionate influence on this planet.

The second thing to keep in mind about "language as we know it" is that language is as sophisticated in all human cultures and is thus truly a trait of the species, not of a certain "civilized" subset of the species. In other words, in this book, "language" is not shorthand for just the languages encoded in newspapers, serving as vehicles of great literature, used on the Internet, and taught badly by Berlitz (for decades, the first sentence in the Berlitz English self-teacher for Spanish speakers was the indispensable and warmly natural "Have you a book?"). One might quite reasonably suppose that a First World culture with tall buildings, cappuccino, and Pokémon would have a grammatically "richer" language, necessary to convey the particular complexities inherent to our treadmill to oblivion, whereas preliterate cultures such as, say, those in the Amazon rain forest would have "simpler" languages for simpler lives. "Bunga bunga bunga!!!!" as the "natives" say in old cartoons.

Ironically, however, if there is any difference along these lines, it is the opposite: the more remote and "primitive" the culture, the more likely the language is to be bristling with constructions and declensions and exceptions and bizarre sounds that leave an English speaker wondering how anyone could actually speak the language without running the risk of a stroke. Meanwhile, many of the hotshot "airport" languages are rather simple in many ways in comparison with the "National Geographic" cultures' languages: English, Spanish, and Japanese grammar are "Romper Room" compared with almost any language spoken by the hunter-gatherers who first inhabited the Americas. In short, one could inform one's friend about giant squid and how they have been encountered in all six thousand of the world's languages with all of the nuance and precision with which we could express these ideas in English.

Thus "language" in this book is not a shorthand designation either for how Westerners talk when wearing their Sunday best or for writing. For our purposes, "language" is perhaps most appropriately symbolized by a rain forest inhabitant speaking while seated near a cooking fire. The "language" that this book is about is the spontaneous and, at base, oral phenomenon that, even in what we might perceive as its homeliest guises, is of a complexity distinguishing our communicative abilities from those of any animal and that worldwide displays a marvelous sophistication not correlative in any way with level of societal development.

We may never know exactly when human language arose. However, as mentioned earlier, deduction suggests that it has most likely existed for about 150,000 years. Archaeological and fossil remains of human beings suggest that some feature possessed by *Homo sapiens* beyond simple brain size was crucial in enabling this species to take over the world. While the brains of earlier species of the genus *Homo,* such as *habilis* and the later, more sophisticated *erectus,* became increasingly larger over the millennia, no cultural development accompanied this increase in brain size. Human existence was typified by *Homo erectus* in northern China, who, in linguist Derek Bickerton's irreplaceable description, "sat for 0.3 million years in the drafty, smoky caves of Zhoukoudian, cooking bats over smoldering embers and waiting for the caves to fill up with their own garbage." Only with *Homo sapiens* do we see an abrupt cultural explosion: symbolic artifacts buried in graves, evidence of nomadic life styles following game instead of maintaining one home base and traveling farther to reach the game when it migrated (which appears to have been typical of Neandertals, who died out in the face of *sapiens*), and, by 35,000 years ago, a major turn in the intricacy of tools.

As Bickerton has argued, all of these things suggest a fundamental transformation in mental ability toward the symbolic kind of cognition which underlies human language. If language arose approximately when *sapiens* did, then a combination of the fossil record and modern comparative genetic analysis can point us to language's time of origin. Specifically, our oldest *Homo sapiens* fossils come from, as it happens, Shirley's homeland, Israel, as well as South Africa, in both cases dated at about 100,000 years old. The hominid fossil record is notoriously fragmentary, such that we may very well find older *sapiens* fossils in the future. Until then, comparative genetic analysis allows us to trace the species back somewhat farther, having repeatedly placed the origins of modern humans in East Africa between 150,000 and 200,000 years ago. Thus it would appear that human language can be traced back at least 150,000 years.

Yet there remains the problem that only just 35,000 years ago do we see the kinds of cultural explosions among human beings that mark them as indisputably "us." The possibility theoretically remains, then, that language did *not* arise right when *sapiens* did, but instead only arose, say, 35,000 years ago.

However, other evidence suggests the earlier date. Especially important here is evidence that human language is to some extent genetically coded. How specific and detailed this innate inheritance may be is controversial (for the innatist view, see Steven Pinker's *The Language Instinct;* for particularly cogent alternative viewpoints, see Terrence W. Deacon's *The Symbolic Species* or the especially lucid *Educating Eve* by Geoffrey Sampson). But various indications suggest that human beings are at least genetically predisposed to acquire and use language. One indication is that damage to specific areas of the brain can have highly particular effects on language ability (one kind of damage leaves people using words without their meanings; other kinds interfere with people's ability to use endings or sometimes *particular types* of endings; etc.). This suggests that language is not entirely just a conditioned skill grafted onto more general aspects of cognition and that our brains have evolved in a direction uniquely suitable to processing language. Another indication is that babies babble spontaneously in all cultures, regardless of whether the culture is predisposed to "goo goo" at them or "teach" them to talk (all are not).

In regard to dating human language, what is important here is that if this genetic instruction or predisposition for language is real, then it must have been created by a mutation. In this light, it is more economical to reconstruct that such a mutation occurred once in the stem population of *Homo sapiens* 150,000-odd years ago and was then passed on to all descendants, rather than emerging at various later times in separate offshoot populations. Indeed, some traits can mutate into existence separately throughout the world, such as the development of eyes or the power of flight. However, if the development of language were susceptible to convergent evolution in this way, we would expect that at least one or two other species had evolved or would be evolving a similar ability, and yet there is not the slightest sign that they ever have or will. No baboon colonies that talk have been smoked out; no mice or ducks or rabbits that converse (much less wear white gloves and pay taxes) have been encountered. This suggests that language was a particularly unique mutation most likely to have occurred once.

This argument becomes even more compelling given that, if language had emerged in separate populations at later dates, then we would expect there to remain pockets of human groups where this mutation had not occurred–or at least for there to be records of such in the past. Yet we know of no such groups, which again points us to reconstructing that the trait arose at *sapiens'* origin, before human groups had split up, the feature then persisting in all of them.

Of course, it may eventually be shown that there is no genetic predisposition for language and that language is indeed an artifactual "graft" onto humanity rather than an innate trait (from my reading of the facts, this conclusion is just as likely to be reached in the future as the discovery of an innate language capability). Yet even here logic would dictate the reconstruction of a single original language: to propose that offshoots from the first group of human beings eventually developed language anew is to presume that this offshoot group had for some reason ceased using language in the past. But given the obvious advantages that language confers on the species, it is extremely unlikely that any human groups have ever cut out talking. Anthropologists have found no such human group in the present, for instance, although cultures do vary in how much they value speaking in general (the Puliyanese of South India barely talk at all after age forty; Danes tend to be on the quiet side; Caribbeans less so; the Roti of East Timor process silence as downright threatening and appear to talk a mile a minute all the time).

Thus the facts available to us at this writing lend themselves most plausibly to the hypothesis that the first human language emerged roughly 150,000 years ago in East Africa. We do not and never will know any words from this first language, nor much of anything about it at all. (In the Epilogue I will discuss the truly tantalizing but ultimately untenable claims that words from this language are reconstructable.) Certainly, there was no way to record this language mechanically, nor was it written (writing of any language would not begin until several tens of thousands of years later, in about 3500 B.C.). Nor do the people who live in East Africa today speak that language: even if there are people living there today descended directly from the original group, their language would by now have "morphed" into one completely different (as we will see that all human language is always in the process of doing, no matter what the conditions); besides, the inhabitants of East Africa today appear to be all or mostly descended from peoples who migrated there subsequently.

What we do know is that what was most likely one original language spread by offshoot populations first to Asia, with one group eventually migrating to Europe, while another spread in two directions: southeastward across Asia down to Australia and northeastward across the Bering Strait to the Americas (mounting evidence suggests that there were also some migrations across the Pacific to the Americas). During these movements, the original language eventually evolved into thousands of others, resulting in the roughly six thousand languages extant today. The process by which one original language has developed into six thousand is a rich and fascinating one, incorporating not only findings from linguistic theory but also geography, history, and sociology. It is this fascinating story that I will share with you in this book.

My aim is to tell a story that has yet to be shared with the general reading public, rather than to contribute by reinforcement to spreading messages about language already treated in accessible sources. As such, in this book you will not find many of the things often covered in books about language. Although you will encounter a dazzling variety of languages, there will be no attempt to provide a family-by-family survey of the world's major languages for its own sake. Various books explore the history of individual words; etymologies will figure in this story only where they are illustrative of a larger point. Similarly, this book will not entail a defense of the legitimacy of slang, an outline of the development of writing systems throughout the world, an exploration of the ways in which language reflects culture, or an exposé of the folly of "blackboard grammar" rules such as the one designating *Billy and me went to the store* "wrong." All of these topics have been treated, in many cases often, by other authors.

Of course, most of those things will make the occasional appearance as we go along. However, the principal intent of this book is to foster a new conception of "language" entirely. This intent springs from something of increasing concern to professional linguists, which becomes increasingly urgent amid the present-day flowering of books and magazines presenting academic findings to the reading public in myriad disciplines. There is a long-standing gulf between how the general public tends to conceive of language and what linguists have discovered about how languages change, relationships between languages and dialects, and how they mix.

This is not the public's "fault," because these concepts are not taught in secondary schools, are generally taught in passing even in undergraduate introductory linguistics courses (which, of course, only a minority of students take), and have been only fitfully disseminated in the form of accessible presentations.

Did you know that according to the instructions for Monopoly, if you own all of a property group (say, the "reds" Kentucky Ave., Indiana Ave., and Illinois Ave.), then even if you haven't built houses on the properties yet, people landing on them are to pay double the base rent? I have never known anyone who observed that rule. In general, almost nobody plays Monopoly straight from the instructions, and almost everybody adds their own rules–in my house, we gave anyone who rolled snake eyes (2 ones) $1,000; some people allow you to build houses on a property before you own the whole set of them; the grand old tradition of putting fees exacted by the Chance and Community Chest cards into a "pot" collected by people who land on "Free Parking" is not in the instructions, which on the contrary actually specify that "a player landing on this space does not receive any money, property or reward of any kind." None of us lose any sleep over our little variations, but most of us know that there is a "real" Monopoly, specified in the small print of that dull, wordy little instruction sheet rarely read and often lost, that we are not ever quite playing.

We are taught, passively but decisively, to think of a language as being like those Monopoly instructions: unquestionably inert and static and "given" from on high, with departures therefrom constituting petty violations of something inherently immutable, a game so eternal and so deeply embedded in the national fabric that it doesn't even have to advertise (no "Pretty sneaky, Sis"-type commercials for Monopoly on television, if you think about it). Sure, we all know that slang changes from decade to decade and even that, as history forces a certain object or concept out of use, the word for it tends to disappear along with it (such as the medieval instrument called a *shawm* or the antique ailment name *neurasthenia,* a vaguely characterized malaise that makes one think of Gilded Age presidents' wives). In the same way, Monopoly houses and hotels used to be made of wood, and through the years a "deluxe" model emerged with extra playing pieces among other things. But, overall, the board will never change; the little man will always wear a top hat, although that sartorial gesture is now seventy years out of date; the car piece will never be changed to a BMW; the rules will always remain the same as they were when the game was released in 1935, and Parker Brothers will presumably never have any reason to revise them.

Yet the truth is that everything about a language is eternally and inherently changeable, not just the slang and the occasional cultural designation, but the very sound and meaning of basic words, and the word order and grammar. For example, seven hundred years ago (when Michael Crichton's *Timeline* takes place), in the language English that I speak and am writing in, *name* was pronounced "NAH-muh" rather than "NEIGHm," *silly* meant "innocent," and double negatives were good grammar. Three thousand years ago, the French language that we know of today was spoken by no one, because it did not yet exist; it was still Latin, which only developed into French through a profound transformation of all of its sounds, sentence structures, and most of its basic word meanings (three Latin words, *de, de,* and *intus* "from," "from," and "inside," eventually squashed together to become a word for "in" pronounced roughly "dong," *dans*). It is even less obvious to us on a day-to-day basis that it is natural for languages to mix to various degrees, such that none of the world's languages are "purebred," all of them having been imprinted to some extent by other languages, at least in regard to vocabulary but just as often all the way down to their very grammars. This is not only a "jungle" affair happening through barter or some other condition that most of us process as "other": a mere *one percent* of the words in English today are not borrowed from other languages.

In short, though we are taught that language is like a copy of Monopoly instructions, language is actually analogous to cloud formations. We look at a cloud formation with full awareness of its inherently transitory nature: we know that if we look up again in an hour, the formation will almost certainly be different and that if it isn't, then this is due to an unusually windless interval that will surely not last long. Language does not change that fast, of course, but it changes just as inevitably and completely over time. Language is an inherently dynamic, rather than static, living entity. One sees or hears that said occasionally, but usually in reference to the inherent liveliness of slang or to the fact that language is used by living beings and rooted in changing cultures. Both of these things are true, but they are only a beginning: language is as changeable an entity as cloud formations even in its mundanest, most "vanilla" aspects such as the words *dog* or *since.* Even when we say any of these things, we are utilizing a system that is eternally mutating, in a slow but inexorable process of becoming a new system entirely, like the lava in one of those lava lamps from the '70s.

For people speaking the language I am writing in a thousand years ago, *dog* was pronounced "DAW-jah" (spelled *docga*) and was a secondary word rather like *fowl* is today; the usual word for *dog* in general was *hund*, which has limped down to us as the now marginal word *hound*. A thousand years ago, in the language called English, *since* was a compound word *siththan* from the words for *after* and *that* and was only used in the chronological "after that" sense of *She has been sad since the day her fish died*; the "because" usage *(He has to have been there since they found his umbrella in the basement)* would only become established five hundred years later. And English isn't special: all six thousand of today's languages have arisen through just this sort of gradual change from the first language spoken more than 100,000 years ago on the savannas of East Africa.

The parallel with the evolution of animals and plants is obvious. The fit is far from perfect. Whereas organisms' evolution is constrained by the central goal of propagating genetic material, languages evolve not with any "goal" to keep themselves going, but simply because it is as inherent for them to evolve as it is for a cloud formation to change. (To those of you who are inclined to object that language evolves strictly to express and preserve culture, I address that issue in Chapter 1.) Yet the process of biological evolution itself is in many ways quite similar to that of flora and fauna. Stephen Jay Gould has told us that evolution is geared not toward progressive "fitness" but toward simply filling available ecological niches. Bacteria, toads, wallabies, and orangutans do not fall on a cline of increasing closeness to God; all four are equally well suited to leading the lives they lead. In the same way, language evolution is not geared toward improvement. Instead, languages change like the lava clump in a lava lamp: always different but at no point differentiable in any qualitative sense from the earlier stage. The process is better termed *transformation* than *evolution*.

Organisms evolve into species and subspecies by mutations. By similar "mutations" and in similar fashion, languages evolve into new languages or, before they have changed to this extent, into "sublanguages," or dialects. Some creatures, like bees, can reproduce both sexually and asexually: a queen bee's fertilized eggs become females and her unfertilized eggs become males. Languages usually reproduce "asexually" by evolving into new ones on their own, but they can also meet one another and yield little-known but rather common language hybrids combining roughly half of one language with half of another. Tiny creatures called tardigrades, which live on wet forest surfaces, can go into suspended animation under dry conditions, pulling in their legs, secreting a protective shell, and suppressing all signs of metabolism. Yet while they're in this condition you can boil them, freeze them, or submerge them in alcohol and they will still come back alive when exposed to water. (Ironically, these critters look like bears, which hibernate in a similar fashion but cannot withstand boiling, etc.)

Languages can similarly be stripped of all but their most fundamental grammatical structures and be used by nonnative speakers for passing communication only (as pidgins–think of Tonto of *The Lone Ranger*); but if conditions arise in which a full language is needed to express any thought, the pidgin can be "awakened" into a full language again (creoles). Some species, like tuataras and horseshoe crabs, find a stable little niche and live on unchanged through the eons, with no need to evolve to fit new conditions. A given dialect of a language has often been assigned a particular static "niche" as the official common coin of a population, codified for use in formal contexts and writing, its lava-lamp transformation retarded–the result is "standard" varieties such as Standard English and Hochdeutsch (Standard German). Flora and fauna can become extinct; languages and dialects do so as well, and, just as we are losing biological species at an alarming rate on our planet, most of the languages that now exist are almost certain to become extinct within this century.

Thus the combination of wonder, injury, jealousy, and rue that little Shirley stirred in me that day in 1970 was due to the contrast between just two of the thousands upon thousands of variations on that one original language that have arisen in the past 150,000 years or so. In the pages that follow, we will take a trip through the natural history of human language and explore how its eternal and inexorable mutability and mixability have transformed the sounds, sentence patterns, and word meanings of one Ur-tongue into six thousand new languages. We are taught from childhood about how art, music, dance, cooking, dress, technology, and even private life began and developed throughout the history of humankind. Here, our guiding question will be a simple one: What happened to the first language?

1

The First Language Morphs into Six Thousand New Ones

I am always a step behind when it comes to technological developments. At the start of my graduate study at Stanford in 1988, I had no idea what "e-mail" meant when I encountered it on a personal data form, but soon discovered that for most of the people in the department, e-mail had largely replaced the telephone, written letters, and memos. It took me about three years to incorporate e-mail into my routine. By 1998, it was the World Wide Web that, for people with computers, had become a norm rather than a marginal toy, first choice for movie listings, personals ads, travel booking, and fact checking. I still use the Web more when I must than as an ingrained habit. My next problem will be cell phones, which by the summer of 1999 became "default" in the United States. It has gotten to the point that saying that I don't have "a cell" lends me, I suspect, the air of a sequestered holdout that we sense in people who do not have VCRs. I'll have given in by the time you read this, but by then I'll probably be among the last people in America not reading e-mail on their wristwatches.

My problem is that I have never been comfortable with change. I have an illogical underlying notion that under normal conditions life stays eternally the same and that changes constitute occasional and disruptive departures from this stable norm. A dinner six years ago will sit in my mind as so recently past that it is unofficially still in the present, whereas the person I was with will barely remember it; when a child I haven't seen for years is now much bigger and more articulate, I have a hard time shaking a sense that some trick has been played. But of course, as everybody but me seems to accept with no trouble, change is not an exception; life *is* change. Kids grow, musical styles change, people keep inventing things, *Seinfeld* goes off, women get pregnant, men go bald.

Most of us are less aware that language, too, *is* change. All human speech varieties are always in a constant process of slow transformation into what eventually will be so different as to be a new language entirely. This change is certainly *influenced* by historical, social, and cultural conditions but is not *caused* by them alone; the change would continue apace even without these things. Human speech transforms itself through time just as vigorously, and even more so, in isolated hunter-gatherer societies where cultural change of any kind has been minimal for millennia. Just as we can understand biology only by being fully aware of the centrality of evolution to how life as we know it arose and will develop, we can truly understand language only by shedding the Monopoly-instructions conception that school inculcates us with and replacing it with a conception of language as a fundamentally mutative phenomenon.

We begin by exploring a basic question: As that initial band of hunter-gatherer *Homo sapiens* migrated northward carrying the first human language with them, what happened to the language? In other words, why doesn't anyone speak it today? Well, the first thing that happened to it is that as time went by, and especially as the original band multiplied and split into offshoots, the first language gradually turned into several thousand different ones. How did this happen, though? How *does* a language change beyond the likes of the coming and going of expressions like "That was hella cool" or isolated words like *thou*? Why didn't the language stay the way it was, with only slang words and expressions differing from place to place? What happened to the first language?

Language Change: Complete Overhaul

The transformative nature of language is as difficult to perceive as the fact that the mountains that look so indestructible to us are gradually eroding, to be replaced by new ones "thrown up" by geological collisions we never seem to see. We can only perceive the changes that occur quickly and frequently enough to fall within a human life span, and thus, just as we are aware of earthquakes and volcanic eruptions, we are all aware that languages change, but mainly on the level of slang.

In New York City in the 1830s, young people of the merchant class were given to saying things like, "I blew up the post office." But the people using this expression were nonviolent people with jobs. "Blowing up" an establishment meant walking in and giving the management a piece of one's mind for a slip-up or discourtesy—"I went in and blew up the post office when I found out they lost that letter." That seems as queer to us today as the corsets and waistcoats the people suffered in as they said it, because slang comes and goes like fashion. "That's for mine!" a flapper might have said in the 1920s to mean exactly what her equivalent in the 1980s would have put as "That rocks!"

This sort of thing, however, is merely the outer layer of the kind of change that all languages undergo, the profundity of which we can only see when we juxtapose a language at two points separated by a good millennium or two. Changes in slang will have been so buried by the turning inside out and upside down of everything that made the language recognizable as itself that only in the intellectual sense are we dealing with one language: instead, we see a language that has evolved into another one.

Here, for example, is a sentence in one human language as spoken in A.D. 1, followed by the same sentence in the language as spoken in A.D. 2000. The sentence itself is quite randomly chosen, likely to be uttered by all of us several times in any given month:

> Admit it, my sisters—the woman hasn't even seen the talking dog!
>
> A.D. 1: Agnoscite, sorōrēs meae—fēmina ne canem loquentem quidem vidit!
>
> A.D. 2000: Admettez-le, mes soeurs—la femme n'a même pas vu le chien qui parle!

Most readers probably recognize that the first sentence is Latin and the second is French. Latin and French are completely different animals for us today, treated in different books, taught in different classes: Latin is laurel crowns and *e pluribus unum,* French is pursed lips and *je ne sais quoi.* Yet French is nothing other than Modern

Latin: Latin as it changed through several centuries into a new language in the area that would become France. We only happen to be able to juxtapose the two stages in the development of this one language because the advent of writing has preserved Latin for our perusal. When Latin arose, French did not yet exist; without Latin, there would never have been anything that could turn into French–in other words, French *is* Latin. When we say that language is always changing, then, what we mean is that the sentence from A.D. 1 gradually morphed, year by year, generation after generation, into the sentence from A.D. 2000.

This kind of change entails the concurrent and interactive progression of a number of processes, which can be broken down into five principal ones. Before we take a look at them, it will help to break down the two sentences. The way linguists do this is to place the translations of individual words underneath.

The Five Faces of Language Change

1. Sound Change: Defining Deviance Downward

Much of the difference between the Latin and the French sentences is due to the fact that in all languages, there is a strong tendency for sounds to erode and disappear over time, especially when the accent does not fall upon them. This is part of what transformed *fēmina* "woman" in the Latin sentence into *femme* in the French one, which is pronounced simply "FAHM." The first syllable of *fēmina* was accented–"FEH-mee-nah"; the other two were not, and over time they weakened and dropped off completely. In real time, we process this kind of erosion as sloppy: to us, *Jeet yet?* is a barefoot version of *Did you eat yet?*, as inevitable but formally unsavory as an unmade bed. But this very process was part of what turned Latin into French, and not "sloppy" French but the toniest *formal* French.

Sounds do not vanish in a heartbeat; at first, there is just a tendency to pronounce the sound less distinctly in casual, running

---

Agnoscite, sorōrēs meae–fēmina ne canem loquentem quidem vidit!
admit sisters my woman not dog talking even saw

Admettez-le, mes soeurs–la femme n'a même pas vu le chien qui parle!
admit-it my sisters the woman not-has even not seen the dog that speaks

speech. What follows is a kind of analogue of "defining deviance downward," a societal trend in which the gradual acceptance of behaviors once considered taboo has the effect of rendering behaviors of the next level of extremity easier to contemplate and fall into ("If smoking pot is no big deal, then why not . . . ?").

A generation that grows up hearing the sound produced less distinctly most of the time gradually comes to take this lesser rendition of the sound as the "default." Meanwhile, however, they, too, follow the general and eternal tendency to pronounce unaccented sounds less distinctly and thus pronounce their "default" version of the sound, already less distinct than the last generation's, even *less* distinctly. The next generation takes *this* muffled sound as "default"; but when they in turn follow the natural tendency to pronounce this sound even *less* distinctly much of the time, this time there is so little left of the sound that to muffle it is to eliminate it completely. Thus, for them, the choice is between making the sound at all and leaving it off completely. Finally comes a generation for whom the "default" is no sound in that position at all.

This erosion has a particularly dramatic effect in that, whereas some sounds in a word serve no particular purpose (the *-ina* of *fēmina*), other sounds are part of suffixes or prefixes that perform important grammatical functions. For example, think of the *-ed* that marks past in English; without this suffix, one does not know from the word whether *walked* is present or past at all. The erosion of prefixes and suffixes like these was particularly central in turning the Latin sentence into the French one. There is a certain tendency for sound change to "go easy on" these prefixes and suffixes to preserve important aspects of the language's machinery. But this is only a tendency, and just as often sound change wreaks its termite-like destruction even on the support beams of a grammar.

For example, the verb *parler* "to speak" in the French sentence descended from a liturgical Latin equivalent *parabulāre* (we will see later why French did not inherit Latin's *loqui*). *Parabulāre* had different forms for all six combinations of person and number in the present tense:

parabulō	"I speak"	*parabulāmus*	"we speak"
parabulās	"you speak"	*parabulātis*	"you (pl.) speak"
parabulat	"he speaks"	*parabulant*	"they speak"

But French *parler* has only three different forms in the *spoken* language: two forms with endings and the other with no ending at all. *Written* French indicates five different endings, but three of them (*-e, -es, -ent*) have long ceased to be pronounced; as in the word *femme*, the spelling preserves a long-lost stage when the erosion had not happened yet. The French child, knowing nothing of writing yet, learns just three forms of *parler*:

je parle	[PARL]	*nous parlons*	[par-LO^{ng}][1]
tu parles	[PARL]	*vous parlez*	[par-LAY]
il parle	[PARL]	*ils parlent*	[PARL]

This difference has an effect beyond the words by themselves. The Latin verb forms are given without the pronouns (*ego* "I," *tu* "you," and so on); the Romans barely needed them, because the endings told them what person and number was intended. In French, however, the pronouns are *de rigueur*, because for four of the forms there is no ending to indicate who is doing the speaking.[2]

This erosion also shaved the case endings from Latin nouns. In Latin, endings conveyed the function of a noun in a sentence: for example, *canem* is the accusative form of the word for "dog," indicating that it serves as the object of the sentence; the nominative

1. This little "ng" is an unavoidably approximate way of signifying that the vowel is nasalized, as in the last sound in François Mitterrand's name, or, for fans of *The Little Mermaid,* the sounds in the immortal Howard Ashman lyric in the scene where the chef is chasing the fish: *"Les poiss**ons**/les poiss**ons**/hee, hee, hee/**hon, hon, hon!**"*

2. There are languages with no person or number endings that *still* don't use any pronouns, such as Japanese, which can be bizarrely telegraphic from the English perspective–*I like Masako* is just "Masako likable"–but this is not the usual case, largely concentrated in various languages of East and Southeast Asia.

Agnoscite, sorōrēs meae–fēmina ne canem loquentem quidem vidit!
admit sisters my woman not dog talking even saw

Admettez-le, mes soeurs–la femme n'a même pas vu le chien qui parle!
admit-it my sisters the woman not-has even not seen the dog that speaks

---

("default" form) was *canis*. Notice that the French word for "dog," *chien*, has no ending signaling that it is the object: there's no "chien-em" or the like; the word is the same no matter how it is used. Because there are no such endings to convey what the object is, French uses a relatively rigid word order: what indicates that something is an object is its being placed after the verb. Latin, still retaining the case endings, had much freer word order, such that our sentence could have also been *Agnoscite, meae sorōrēs–fēmina quidem ne vidit canem loquentem* or *Agnoscite, sorōrēs meae–canem loquentem fēmina ne quidem vidit.*

Thus the gradual erosion of a language's sounds not only slowly renders all words into new ones often barely recognizable as descendants of the originals. It also transforms the very grammar of the language.

But if all that sound change consisted of was erosion, then presumably all language would long ago have worn down into a mouthful of dust. Fortunately, sounds also transform into new ones. Imagine Moe the Bartender on *The Simpsons* or Moe of The Three Stooges saying, "Shut up!" What they say is actually essentially "Shaddap," with the *a*'s quite close to the vowel in *cat*. Where this transformation began was when speakers developed a variation of the *uh* sound that had a slight hint of the *a* in *cat* mixed in, but it was not as close to that *a* sound as the current "Shaddap" version. These people thus alternated between two versions of the sound–*uh* and what we will call "*uh*-plus"–just as people can alternate between a full pronunciation of an unaccented sound and a muffled one. This was a manifestation of a general instability in vowels–in all languages, they tend to gradually mutate into different ones as time goes by.

What followed was more "defining deviance downward." In these people's minds, "*uh*-plus" was an "alternate" version of the "real" *uh* sound. But a child might hear his elders' "*uh*-plus" sound not as a deviation from the "real" sound but, instead, as the "default" version of the sound, especially if the elders tended to use "*uh*-plus" more often than *uh*. But the child is a human being, too, and will develop his own "alternate" version of his default vowel. In accord with the trend set by the people around him, that alternate will lean a bit farther in the direction of *a* in *cat*. Then *his* children interpret *that* "version" as the default and develop their own alternate even farther in the *a*-in-*cat* direction.

The slight difference between the *uh* vowels of one generation and another is largely undetectable (except to linguists doing painstaking recording and statistical analysis), but the cumulative effect of a process like this is "Shaddap" or "Caditout" for "Cut it out"–the sound change does not usually affect only one word but spreads to most or all instances of that sound in the whole language. If you think about it, Moe the Bartender's *but* leans toward "bat"–when you wrap your head around hearing it for itself rather than perceiving it as "a version of *but*"; his *What?* is kind of like a "wat" rhyming with "rat." The sound has traveled so far from *uh* that the only reason we now sense it as "a version of *uh*" is because English spelling preserves what the language happened to be like before the change started. Moe's son (if the bartender or the Stooge had one, Lord forbid) might well write the expression as "Shaddap" before being taught how to spell–to him, he *is* saying the *a* in *cat*.

We see this kind of sound change in the pathway from Latin to French as well. Latin had "FEH-mee-nah" for *woman*, but despite the misleading spelling, French has not "FEHM" but "FAHM." There are myriad possible paths from one vowel to another as a language changes, of which *uh* to *a* in *cat* is but one: from Latin to French, many *eh* sounds changed to *ah* by a gradual transformative process similar to the one that produced "Shaddap" from *Shut up*. The French spelling *femme* is today as out of step with how the word is pronounced as the English *tough*; *femme* is merely a relic of what earlier French was like before the vowel had changed. In another transformation, the *ah* of *canem* [KAH-nem] became the *eh* of *chien* [SHYEH^{ng}], a general change also visible in pairs like Latin *cārus* [KAH-rus] "dear" and the French *cher* (pronounced like Chastity Bono's mother's name).

2. Extension: Grammar Gets a Virus

The second process that changes languages into new ones worldwide is a tendency for some-time patterns in a grammar to generalize into exceptionless across-the-board rules. For example, if we

---

Agnoscite, sorōrēs meae–fēmina ne canem loquentem quidem vidit!
admit sisters my woman not dog talking even saw

Admettez-le, mes soeurs–la femme n'a même pas vu le chien qui parle!
admit-it my sisters the woman not-has even not seen the dog that speaks

wanted to make the words for *woman* and *sister* plural in Latin, then we needed different rules. *Sorōrēs* "sisters" is a plural form with *-es*, but *fēmina* in the nominative plural was *fēminae*. These words belonged to different classes of noun, whose sets of endings differed. There was, for example, another set of nouns whose (nominative) plural ending was *-i*: *dominus* "master" was *domini* in the plural.

As Latin's endings wore away while Latin was transforming into French, only one of the three plural endings was left behind, and speakers began to use this ending with all nouns instead of only those of certain noun classes. The plural of *femme* is *femmes*, using the *-s* marker that in Latin was used only with certain nouns, of which *fēmina* was not one. The plural of the French words that developed from *dominus*–*dom* and *don*–both pluralize with *-s* as *doms* and *dons,* and so on. The *-s* ending spread like a virus and has now taken over the whole organism.

This happened in English, too: English began as a language like Latin in which nouns had case endings and fell into various classes with differing endings. The plural of *fox* was *foxas*, but the plural of *tunge* "tongue" was *tungan*, the plural of *waeter* "water" was the same as the singular, whereas the plural of *bōc* "book" required not an ending but a change in the vowel to *bēc.* As the endings wore off one by one, just as they had in "Fratin" across the English Channel, one of the plural endings, the *-s*, took over: now we have not only foxes, but tongues, waters, and books. If this hadn't happened, then the plural of *book* would be *beek*! Instead, only the strayest of remnants of things like this lurk in the language, such as *mice* and *brethren*.

3. The Expressiveness Cycle: "The Bass from Hell"

In 1988, I was in a musical in which one man in the chorus regrettably could not be said to have a talent for singing. Theater is an inherently gossipy affair, and the chorus members started referring to him behind his back as "the bass from hell." This was the first time I had heard this expression, and at the time it was quite funny; hearing it, one got a mental picture of a man emerging from the pits of Hades to torture us with a preternaturally unpleasant singing voice.

Nowadays, though, "from hell" is used by young people to refer to anything even slightly noisome, such as a cloudburst or dull weekend: "Like, I had to drive him the extra ten miles in traffic–it was so from hell." In the '90s, the expression gradually stopped arousing a literal vision of the underworld and became a general indication that something was kind of lousy.

Terrible has had a similar history over a longer period. Originally, *terrible* referred to truly horrifying things like the movie *Showgirls* or giant squid–things that leave you sincerely wondering how they could come into being on the earth as we know it. Through the years, however, it gradually came to be used for phenomena less and less grisly, such that today we casually apply *terrible* to things like unsavory meals or unsightly architecture. But this kind of thing cannot be predicted, only explained afterward; in many other languages, the word for *terrible* retains its original force, whereas other words are sent down the treadmill to semantic oblivion.

Russian is such a case, and it creates little ripples of semantic ambiguity in translations. In his translation of the short piece "Father's Butterflies" by his father, Vladimir Nabokov, Dmitri Nabokov has a neat little reference to "the terrible turtles who direct learned journals." Yet *terrible* as used in English rings a little weak here in itself; the soul of this phrase is Russian, in which the word *strašnyj* connotes "terrible" in the sense the word once had in English, "terrifying" like Maurice Sendak's "Wild Things." The Russian connotation is the nightmarish iniquity of mediocre thinkers holding considerable power over one's output and career path, whereas taken as "English," *terrible*, having been diluted into referring to things like slow traffic, has a less cosmic ring and sounds more like a passing disparagement of the editors' talents. Dmitri Nabokov's translation takes advantage of the contrast in the semantic evolutions of *terrible* and *strašnyj* to convey a feeling of Russian through English.

Indeed, part of the reason we needed a "from hell" is because words like *terrible* have worn down so: all languages constantly create expressive usages of words or phrases that gradually wear down

---

Agnoscite, sorōrēs meae–fēmina ne canem loquentem quidem vidit!
admit sisters my woman not dog talking even saw

Admettez-le, mes soeurs–la femme n'a même pas vu le chien qui parle!
admit-it my sisters the woman not-has even not seen the dog that speaks

---

in force, like old jokes. Manifestations of this process are responsible for much of the difference between our two sentences.

Notice, for example, that Latin has no definite articles for "the woman" or "the talking dog," but French has *le chien* and *la femme.* Latin was like a great many of the world's languages, such as Russian and Chinese, in having no words for our *a* and *the*–English and its relatives are actually rather odd as languages go in having words serving those functions. French's definite articles trace back to what began in Latin as words for "that": *illa fēmina* meant "that woman" (that is, Bill Clinton's pet name for Monica Lewinsky). Words for *that* serve to explicitly point out an object or concept and distinguish it from others, just as Clinton accompanied his denial with a jabbing of his pointer finger (in linguistics they are even called *deictics,* from the Greek for "to point"). This is the kind of expressive force that tends to diminish in a language through time, just as the meaning of *terrible* in English did. As Latin became French, the connotation of *illa* (and its variants *ille, illud,* etc.) weakened into that of English *the,* which lends a noun a *certain* specification–*I saw the man* implies that the man has been talked about before, whereas *I saw a man* introduces the man into the conversation for the first time–but with nothing approaching the explicit, finger-pointing force of *that.* In fact, whereas distinguishing something as *that* is an occasional thing (such as when Kenneth Starr comes after you), just about everything you mention in the course of a conversation has been specified before and is thus ripe for marking with a *the.* Therefore what began as the explicit Latin *illa* became, with some erosion of sounds into *la,* the definite article that now must be used with all prespecified feminine nouns in French.

French is even more addicted to definite articles than is English: whereas we say *Milk is white,* the French, since milk is after all something known already to all of us, say *Le lait est blanc* "The milk is white" even when referring not to "that" milk, such as that which we've been talking about going to the refrigerator to drink up, but just milk as a substance in general. If any French person has ever actually said, "Let us make ze love," as Pépé le Pew did, this was why.

Loss of expressivity had a particularly dramatic effect on the difference between these sentences in regard to negation. Notice that where Latin had just one marker of negation (in our case, *ne*), French has two: "I am not walking" in French is *Je ne marche pas,* where both *ne* and *pas* serve to indicate "notness."

The reason for this is the same one that created English's "terrible" scrambled eggs. In earliest French, there was only one negative marker, *ne*: *Il ne marche* "He is not walking." However, you could reinforce the negation with various expressions conveying the meaning "not one bit," such as:

mie "crumb" *il ne mange* "he doesn't eat" versus

il ne mange mie "he doesn't eat a crumb"

goutte "drop" *il ne boit* "he doesn't drink" versus

il ne boit goutte "he doesn't drink a drop"

Along these lines, for walking, you would use *pas*, the word for "step":

pas "step" *il ne marche* "he doesn't walk" versus

il ne marche pas "he doesn't walk a step"[3]

As time passed, these expressions began to lose their snap just as "from hell" and *terrible* have lost theirs. Eventually, they no longer conveyed any more forceful a negation than using *ne* alone, just as, today, calling something *terrible* is often not appreciably stronger a condemnation than calling it "bad." As a result, most of these double-stuff expressions fell out of use. The one with *pas*, however, hung around but underwent a transformation manifesting its faded force. *Pas*, like the other markers, no longer added any substantial force to a negated sentence: just as "Traffic

3. If those of you who know French are wondering why it wasn't *il ne marche **un** pas*, it's because indefinite articles had not become obligatory in such cases yet, just as definite articles still retained a degree of their "thatness" and were not yet used in the *Le lait est blanc* cases.

Agnoscite,	sorōrēs	meae–	fēmina	ne	canem	loquentem	quidem	vidit!
admit	sisters	my	woman	not	dog	talking	even	saw

Admettez-le,	mes	soeurs–	la	femme	n'a	même	pas	vu	le	chien	qui parle!
admit-it	my	sisters	the	woman	not-has	even	not	seen	the	dog	that speaks

was bad" and "Traffic was from hell" now mean the same thing, *il ne marche* and *il ne marche pas* meant the same thing, simply "He doesn't walk." With *pas* no longer contributing any force to the negation, the original literal sense of *pas*–"step," and thus "not one bit"–no longer had any logical connection with the meaning of such sentences. For this reason, speakers gradually ceased processing *pas* in these sentences as connoting any concrete concept at all, instead reinterpreting it as just something one must use when negating a sentence involving movement–in other words, as simply a negator just like *ne* (who said you only had to have one negator word?).

Once *pas* had been reinterpreted in "movement" sentences as simply meaning "not" just like *ne*, it began to spread to verbs beyond *marcher* "to walk" and others having to do with locomotion. This would have been impossible when *pas* was still processed as literally indicating "step": you can't drink or eat a step, for example. But as a simple negator just like *ne*, *pas* was now compatible with any verb, since, as Clinton taught us in 1998, any action is potentially negatable. By the 1500s, *pas* extended, virus style, to usage with all verbs: *il ne mange pas* "he doesn't eat," *il ne boit pas* "he doesn't drink," *il ne parle pas* "he doesn't speak," and so on. Meanwhile, whereas at one point one could use either *ne* alone or use it with *pas*, the use of *pas* was so common that it became a habit and, gradually, like so many habits in life, the rule.

Thus a word that began as a concrete term for "step" (and is still used that way elsewhere in the language–*pas de deux* "two-step") is now just a piece of grammar like our *not*. This is an extremely common process of language change and is called *grammaticalization*. That is, what began as an actual word with a concrete meaning became a word whose only function is to express an aspect of grammar–you cannot hold, caress, execute, burnish, or eat a "no."

4. Rebracketing: The Story of Gladly, the Cross-Eyed Bear

My mother used to recall how, when her church choir would sing the hymn "Gladly the Cross I'd Bear" when she was a little girl, she thought they were singing about a little storybook bear named

Gladly, who was afflicted with an ocular misalignment. The problem here was my mother misassigning functions to the words: the adjective *cross-eyed* instead of *cross I'd* and the animal *bear* instead of the verb *bear.* Language change is driven in large part by misassignments of a similar nature, although on a less fantastical scale.

It starts with little stuff. Did you ever wonder what the *nick* in *nickname* was? What's "nick" about the name? As it happens, the word began as *ekename;* in earlier English, *eke* meant "also." Now that made sense–your "also" name. Through time, however, because the word was used so often after *an–an ekename*–people began to interpret the *n* in *an* as the first letter of the following word. Hence *a nickname.*

What had occurred is a "rebracketing": what began as [an] [ekename] became [a] [nickname]. *Apron* began as *napron,* borrowed from the French word *naperron* for "napkin." Through the same process that created *nickname, a napron* became *an apron.* The French pulled this sort of thing as well: when they gave us their word for a fruit I've never understood why everybody likes, *orange,* its original source had been a Hindi word *nārangī.* In this case, the word gave up its initial *n* to the article preceding it, and hence *nārangī* became our *orange* (that's why *orange* is still *naranja* in Spanish, which didn't amputate the poor word in the way the French did).

This process often combines with sound change to create one word out of what began as two or more. In early England, a *reeve* was an estate manager. The reeve of a county, or shire, was the shire reeve, or *scīr gerēfa* in Old English. Through time, *gerēfa,* because it did not have the accent, began to be pronounced less distinctly, just as Latin case endings did, and gradually became processed as just a sequence of sounds appended to the first word. The eventual result, with further sound changes, was our *sheriff:* [scīr] [gerēfa] to [sheriff]. Did you ever wonder just why that jolly old man is called "Santa Claus"? How many people do you meet at parties named Santa or even read about in the past? The name began in Dutch as *Sant Heer Niclaes,* "Saint Mr. Nicholas"–in other

---

Agnoscite, sorōrēs meae–fēmina ne canem loquentem quidem vidit!
admit sisters my woman not dog talking even saw

Admettez-le, mes soeurs–la femme n'a même pas vu le chien qui parle!
admit-it my sisters the woman not-has even not seen the dog that speaks

---

words, St. Nicholas. Pronounced as often as it was, it shortened to *Santerclaes* and eventually was reinterpreted in English as "Santa Claus," so far from its origin that we do not even perceive its relationship with its alternate "Saint Nick." [Sant] [Heer] [Niclaes] became [Santa] [Claus].

From *nickname* to Santa Claus, English, like all languages, is full of words that are not descended in a "pure" state from older equivalents but have bits of other words encrusted on them or started as two, three, or four words, a natural reinterpretation of the word boundaries having become common enough to become generally accepted. This also happens on a larger scale–natural reinterpretations of the role that words play in whole sentences create new sentence structures. For instance, note that, whereas Latin in our sample sentence conveys "has seen" with one verb *vidit*, French has a construction like English's, with *a vu* meaning "has seen," which is called a *perfect* construction. Latin did not have a perfect construction with *have* like this at first and only developed it through a gradual reinterpretation.

As Latin began to become French (and other Romance languages like Spanish and Italian), a new construction arose using the verb for *have* with a past participle. At first, though, the Latin construction didn't have the *meaning* of our *have*-perfect. In saying

Eam habeō vīsam.
her I-have seen

a Late Latin speaker was using *have* in its full meaning of "possess." What the speaker meant was "I possess her in the state of having been seen." This isn't as bizarre a sentiment to express as it may first appear–an equivalent in English is when we say *But I already have the dress sewn up–why not wear it?* Just as we have the dress in the state of having been sewn up, the Roman had the woman in the state of having been seen–that is, "She is in my sight." This difference in meaning was reflected in how the case endings worked: *Eam* habeō *vīsam*. The ending of the verb form for "seen" agrees with *eam* "her" because it describes something about "her"–what you have is her, all nice and seen, just like you might have a dress, all nice and sewn up. *Casa blanca* "white house"–*eam vīsam* "her (all nice and) seen."

We are used to objects coming after the verb, and so *eam* coming before the verb even though it is part of the object throws us:

perhaps the meaning of the sentence is clearer if we make the word order more like ours (legal in Latin anyway because the endings tell you so much): *Habeō eam vīsam.*

However, if you have something in the state of having been seen, then this implies that in the recent past you saw it–just as if you have a dress nice and sewn up, then it pretty much follows that in the recent past you sewed it. That kind of looming implication has a way of transforming constructions in languages into new ones. The assumption that something you have in your sight is something you came to see in the recent past led to a gradual reinterpretation of the sentence as meaning just that, that you have seen her, just as we would mean it. As part of this transformation, *have* evolved from concretely meaning "to possess" to just indicating the pastness. *Have*, then, went from being a good old-fashioned "real" verb into being a mere helping verb situating "to see" in time–another instance of *grammaticalization* like the evolution of *pas* in French into a negative marker.

Under this new interpretation, the sentence was no longer about having "her in the state of being seen." Instead, it was about you having seen her–in other words, *see* was now *taking* an object instead of being *part* of the object "her in the state of being seen." As such, it no longer made sense for *vīsam* to be marked as an object along with *eam*. Accordingly, speakers gradually stopped marking *vīsam* with the ending agreeing with *eam* and instead gave it the default participial ending *-um*, which agreed with nothing: *eam habeō vīsum.*

Here, then, was a rebracketing of the type we saw with *nickname*, but on a larger scale. With the words that constituted the object in brackets, the expression began as:

habeō [vīsam eam] and evolved into habeō vīsum [eam]
OBJECT OBJECT

Agnoscite, sorōrēs meae–fēmina ne canem loquentem quidem vidit!
admit sisters my woman not dog talking even saw

Admettez-le, mes soeurs–la femme n'a même pas vu le chien qui parle!
admit-it my sisters the woman not-has even not seen the dog that speaks

---

It is the third person singular version, *habet vīsum* "has seen," that further developed, through erosions of sounds, into the *a vu* in our French *La femme n'a même pas vu.*

This kind of reshuffling occurs with various constructions in all languages at all times and is obviously a long way from the rise and fall of little expressions like *I blew up the post office.* This kind of reinterpretation of the warp and woof of a language gradually turns it into a new one.

5. Semantic Change: Making Love to Ginger Rogers

"He made love to me," pouts Ginger Rogers in a strange little moment in the 1935 movie *Top Hat,* explaining to her friend why she feels misled by Fred Astaire. That line sometimes shocks viewers into supposing that the Rogers character meant what we would mean by that today, but *Top Hat* was made too late for the line to qualify as a bit of "Pre-Code Hollywood" realism. In fact in 1935, "make love" could still signify any degree of physical involvement, most often just kissing, which the storyline makes pretty clear is all Ginger meant.

Make love, then, today has a *narrower* meaning than it did even recently, and not only expressions but single words undergo processes of narrowing and broadening through the ages, playing a part in creating a French out of a Latin. Recall from the Introduction that, in English, *hund* referred to any old dog, whereas *dog* (originally *docga*) began as the word for a mysterious breed of apparently large dogs. *Hund*, its thunder stolen, then narrowed into *hound*, referring to hunting dogs. *Dog* gradually broadened in its meaning to refer to all dogs as *hund* once had. This complementary narrowing and broadening often results in the word for even a basic object changing in a language through the years. Broadening, for example, is the process that gave French *parler* for "to talk" instead of Latin's *loqui* (with its present participial form *loquentem* in the sentence). What began as a liturgical verb *parabulare* gradually broadened into referring to speaking in general, while *loqui* itself eventually disappeared entirely, a *hund* with its tail between its legs.

Sometimes a word's meaning simply drifts aimlessly, with each step following plausibly from the last, but the difference between the earliest reconstructable meaning and the most recent one having become so vast as to completely obscure any historical relationship. In Old English, the word that became *silly* meant "blessed." Just as wanting to do something implies that one will do it, blessedness implies innocence. That kind of implication led people to gradually incorporate innocence into their conception of the word, and through time innocence ended up becoming the main connotation rather than the "definition 2" one, just as one sound gradually becomes another one through shades of the new sound gradually encroaching on the original one. Thus, by the Middle Ages, *silly* meant "innocent": about 1400, we find sentences such as *Cely art thou, hooli virgyne marie.* If one is innocent, one is deserving of compassion, and this was the next meaning of the word (a 1470 statement: *Sely Scotland, that of helpe has gret neide*), but because the deserving of compassion has a way of implying weakness, before long the meaning of *silly* was "weak" (1633: *Thou onely art The mightie God, but I a sillie worm*). From here it was a short step to "simple" or "ignorant," and finally *silly* came to mean "foolish"–having begun meaning "sanctified by God"!

Semantic drift has an especially visible effect on combinations of roots and prefixes or suffixes, and this effect, too, creates important differences between a language and the one it turns into. Our French sentence's *admettez-le* "admit it" is a good example of this kind of development. *Admettre* is composed of *ad-*, a preservation of the full word or prefix for "to" in Latin, and *mettre* "to put." This verb had an ancestor in Latin, but this Latin verb *admittere* did not mean "to confess," which was conveyed with other verbs like *agnoscere.* Instead, the main meaning of *admittere* was the literal one of "putting or sending into." The "confess" meaning of its French descendant arrived gradually by the same kind of hanging implication that created French's *have*-perfect: to admit something is to give it entrée into–that is, "put it into"–your acknowledged awareness.

Agnoscite, sorōrēs meae–fēmina ne canem loquentem quidem vidit!
admit sisters my woman not dog talking even saw

Admettez-le, mes soeurs–la femme n'a même pas vu le chien qui parle!
admit-it my sisters the woman not-has even not seen the dog that speaks

The Ghost in the Machine: Language Change in Your Life

All five of the processes of change–sound change, extension, the evolution of concrete words into pieces of grammar, rebracketing, and semantic change–are as natural to language as photosynthesis is to plants and breathing is to animals. I have used Latin, French, and English as examples because they are represented by such a rich written record and because these are languages we are likely to be familiar with. Yet it must be clear that these selfsame processes are and have always been at work in every language on earth: Thai, Navajo, Persian, Arabic, Nepali, Malay, Fijian, Swahili, Chinese, Zulu, Berber. These processes of change are not only why French is so different from its parent Latin, but also why all six thousand human languages differ to an equally vast degree from their single East African ancestor.

Thus, when in *Bram Stoker's Dracula* (1992), Francis Ford Coppola was assiduous enough to have Vlad the Impaler speak Romanian, this was historically inaccurate in a nitpicking sense. The effort that Gary Oldman went to to learn his Romanian lines was commendable, and in going to the trouble of arranging this, Coppola validated his status as one of America's great artists. But technically, since Vlad lived in the 1400s, what he would have spoken would have been the Romanian of six hundred years ago, which would have been quite different from modern Romanian, just as the English of the 1400s would be only fitfully comprehensible to us today. This very matter is minimized in Mark Twain's *A Connecticut Yankee in King Arthur's Court,* where the language problem would have been utterly hopeless, since Old English might as well be German to the Modern English speaker.

More recently, Michael Crichton got closer to the mark in his *Timeline,* in which graduate students travel back in time to a region of southern France owned by English nobles in the 1300s. The students are fitted with miniature machine translators fitted into their ears that allow them to understand the English of the people they meet, without which they would be at a crippling linguistic disadvantage. Yet, if we may split some hairs, even Crichton, who definitely does his homework in crafting his sci-fi pastries, slips up with the details here. The local tongue of the region is French's close relative Occitan, one of the lesser-known Romance languages, which a century before the story had been the language of the strolling minstrel troubadors, in that guise known as Provençal. Crichton gives one of the students an advantage in being a Middle Ages buff who has taught himself to speak Old English and Middle French. The student has also taught himself Occitan, which comes in handy in encounters with various knights and locals. However, Crichton apparently subscribes to the easy misimpression that "indigenous" languages do not evolve, "quaintly" conserving their form along with old dances, songs, and recipes. In fact, Occitan, like any language, has evolved through the centuries just as English and French have; any language changes constantly whether or not it happens to become the vehicle of a "developing" civilization. The linguist distinguishes between Old and Modern Occitan, for example, and this means that the Occitan that Crichton's student learned would be of the same limited use in the 1300s as Modern English or Modern French.[4]

Whoever made up the old joke about the tourist in Rome had it right–a woman travels to Rome with a Latin phrase book, assuming that Romans still speak the Roman language; she asks a local for directions in Latin and the man says, "Ah, I see it's been a while since your last visit!"[5]

Miami is a Native American language spoken not in Florida but in Indiana and nearby regions. It no longer has native speakers.

---

4. As delightful as *Timeline* is, there are other things that the linguist is obliged to just let pass. Crichton has the students versed before their trip in "Old English," when what was spoken in the 1300s, and what he deftly portrays the locals as speaking, is Middle English, much closer to our speech, such that the characters can "get the hang of" talking with people with practice (*methinks, forsooth,* and so on), whereas Old English would have been utterly opaque. If he is diligent enough to have the characters confronted with Middle rather than Old French, then why not Middle English? Crichton also has a character declare that no twentieth-century person could learn Occitan natively because it is "dead," when in fact it is still spoken by fairly large numbers of people in southern France and is the vehicle of a growing literature. There are even Occitan self-teaching books, which would obviously be of limited commercial application if no one actually spoke it! (*Hittite in Forty Lessons*–fluency guaranteed or your money back!) Occitan, however, is indeed threatened in the long term, because, under pressure from French, increasingly fewer young people are learning it.

5. The punch line is always delivered with an inflection (and phraseology) that suggests that the local is for some reason British, but I suppose we're not supposed to look too deeply into this.

French missionaries learned the language starting in the 1600s, however, and recorded a great deal of data in it. In the 1900s, the last generation of native Miami speakers were given this three-centuries-old information from their language and spontaneously noted that it was "old Miami," reminding them of the way they as children had heard old people speaking the language. In other words, Miami, spoken by preliterate hunter-gatherers when whites encountered them, like all languages had always been in a state of transformation, and the only thing that arrested this transformation was its death.

Yeah, but Where Do You Draw the Line? Language Change Is a Continuum Phenomenon

At a party, even if you don't know what a group of people are talking about, you can almost always ease your way into any conversation by simply interjecting at a suitable pause "But where do you draw the line?" It's a general observation that qualifies as an intelligent perspective on almost any conceivable topic–a movie's fidelity to the novel, the domain of independent prosecutors in Washington, school vouchers, the censoring of the orgy scene in *Eyes Wide Shut*–the reason being that almost anything interesting in life is a continuum phenomenon. Our tendency is to put things in pigeonholes and seek binary oppositions: we exalt Linnaeus and physics while treating "fuzzy logic" as a trendy novelty. Yet, in real life, as often as not, the phenomena we observe on our planet are nondiscrete: clines are everywhere.

This is resoundingly true of linguistic phenomena. Language change, for example, is an inherently gradual process, in which any dividing lines that we draw are terminological conveniences at best. People in France did not wake up one morning to find that the language they were speaking was now French instead of Latin anymore than we one day declare that our puppy has become a dog: Latin evolved through "Fratin" into French just as red evolves through purple into blue.

Thus there were centuries during which local speech in what is now France was something that now looks to us like precisely what it was, something somewhere between Latin and French, or "Fratin."[6] Our first attestation of this speech comes from A.D. 842,

6. *Fratin* is not a conventionally accepted linguistic term.

in the form of the Oaths of Strasbourg, when two grandsons of Charlemagne taking an oath had to concede that the local "Latin" was now too different from Classical Latin itself for common people to even pretend to comprehend. What the people were speaking is preserved in this document, as in this sample: *Pro Deo amur et pro christian poblo . . . in quant Deus savir et podir me dunat . . .* ("For the love of God and the Christian people . . . to the extent that God gives me knowledge and power . . .").

In having lost some Latin case endings (note, for example, *christian* devoid of such an ending) and in using infinitive verb forms as nouns as with *savir* "to know" meaning *knowledge* (Latin would have used a noun, *scientia,* for *knowledge*), we are clearly not in Kansas anymore as far as Latin is concerned. But then note that we see *Deo* first and then *Deus*–some case endings still remained at this stage; the placement of the object *savir et podir* before the verb is typical of Latin's free word order but foreign to the French soon to emerge; and the *-at* ending for the third person singular would disappear by the time the language became French.

The earliest speech known as English is so different from the modern language that it is as foreign to us as German: here is the Lord's Prayer in Old English of A.D. 1000:

> Fæder ūre, thū the eart on heofonum, sī thīn nama gehālgod . . .
> Urne gedæghwāmlīcan hlāf syle ūs tō dæg.

By 1300, the language is recognizable as a variety of what I am writing in:

> Fader oure that is ī heuen, blessid be thi name . . . Oure ilk day bred gif us to day.

The only thing that throws us here is *ilk*: it meant "each." Yet listening to this stage of English would have been a rather narcotic experience; most of the time we would manage to get the gist at best. The gulf is clearer without the crutch of a well-known passage; here, for example, is a snippet from *The Canterbury Tales:*

> At Alisandre he was whan it was wonne;
> Ful ofte time he hadde the boord bigonne
> Aboven alle nacions in Pruce;

> In Lettou had he reised, and in Ruce,
> No Cristen man so ofte of his degree;
> In Gernade at the sege eek hadde he be . . . .

The boord bigonne meant "headed the table"; *boord* survives in the meaning of "table" only in the set expression *room and board*, with the connection between *board* and *table* so lost to us that as a child I thought the expression meant that you got a room with just a board in it to sleep on (was it only me who thought that?). *Reised* is "traveled"; German retains the root as *reisen*. *Sege* is just a spelling matter (it meant "siege"), but *eek* is the *eke* that meant *also* in *nickname*'s progenitor *ekename*. *Pruce*, *Lettou*, and *Ruce* were Prussia, Lithuania, and Russia. Now imagine trying to get through a whole dinner of things like this.

But three hundred years later, the Lord's Prayer was:

> Our father who art in heaven, hallowed be thy Name . . . Give us this day our daily bread.

We tend to think of this as Modern English of a "high" or "biblical" style, but it is properly an earlier stage of English, which religious tradition has preserved in state while the spoken language has continued to evolve. Four hundred years later, in Standard English, the prayer is:

> Our father who is in heaven, may your name be hallowed . . . Give us our daily bread today.

There was no discrete dividing line between Old English, which as spoken could have taken days to even recognize as related to what we speak if we never saw it on paper, and "English" as we know it.

Shakespeare, as it happens, writing as the 1500s became the 1600s, wrote in a period when English was just becoming what we would recognize as "our language." This particular place that Shakespearean English occupies on the English language timeline is what makes his language such a challenge for most of us to process when spoken in live performance. In *Hamlet,* for example, Polonius advises:

> Neither a borrower nor a lender be,
> For loan oft loses both itself and friend,
> And borrowing dulls the edge of husbandry.

To us, husbandry is a matter of raising livestock, and thus one must either just let this observation of Polonius go by or ponder what consistent connection borrowing money from a friend has with raising meat. The problem is that Shakespeare catches *husbandry* at a stage in its semantic evolution now past: *husband* began meaning "manager of the house," which then took two paths of evolution through the centuries. *Husband* itself narrowed into meaning specifically "spouse of the woman of the house"; meanwhile, *husbandry* continued to pertain to running the house itself but narrowed in two directions. One direction was toward meaning "raising livestock," which persists, albeit rather marginally, in Modern English. The other meaning arose by looming implication: since running a household presumably entails trying to get the most bang out of one's buck as possible, gradually *husbandry* narrowed to mean "thrift." This is what Polonius means, and the later disappearance of that use of *husbandry* now leaves a sort of jolt of static in Polonius's speech. Since Shakespearean text is sprinkled with things like this, most of us are accustomed to hearing Shakespeare rather like a poorly tuned in radio station.

Yet because we can usually get Shakespeare's basic meaning, the language is "English" to us, and the magnificence of the prose, though somewhat obscure to us without training, discourages us from considering "translating" it for modern audiences as we do *The Canterbury Tales* or *Beowulf,* because to do so would smack of defacing masterworks. Ironically, the continuum nature of language change in this case has it that foreigners have the privilege of being able to take in Shakespeare as easily as we process Shaw. Any translation inherently compromises the original; thus a translator can translate Shakespeare into the *modern* stage of the translator's language, since translating it into that language as it existed five hundred years ago would entail the same degree of compromise of the original in any case. Here is Polonius's advice from six translations of *Hamlet;* note that in whichever of these languages you might be familiar with, the language is more readily processible to the modern speaker of that language than the English is to a native English speaker, especially in regard to *husbandry*, which is usually translated in other languages with simply the modern word for "thrift":

French
Ne prête ni l'emprunte;
Car souvent, par un prêt, l'on perd et l'argent et l'ami;
Quant à l'emprunt, il émousse le sens de l'économie.

Spanish
Procura no dar ni pedir prestado a nadie;
Porque él que presta suele perder a un tiempo el dinero y el amigo,
Y él que se acostumbra a pedir prestado falta al espíritu de economía y buen orden que nos es tan útil.

Italian
Non far debiti e non prestar denaro;
Perché un prestito spesso perde sé stesso e l'amico,
E il far debiti fa perdere il filo all'economia.

German
Sei weder ein Leiher noch ein Borger;
Denn durch Leihen richtet man oft sich selbst und seinen Freund zu Grunde;
Und Borgen untergräbt einer guten Haushaltung.

Russian
V dolg ne beri i vzajmy ne davaj;
Legko i ssudu poterjat' i druga,
A zajmy tupjat lezvejo xozjajstva.

Hebrew
Al na the love, v'al the malve;
Ki lamalve yovdu gam khesef gam yadid,
v'halove–sofo yigda mate lakhmehu.

The slow pace of language change has two effects. One is to condition the illusion that a language is a static system. The other is that a new stage in the change coexists with the older stage for a while before taking over: for example, Latin case endings dropped away slowly through time, not all at once; as we saw, "Fratin" in A.D. 842 still had some. The combined effect is that, where some evidence of language transformation peeks out at us, we process it as "a mistake." There is no indictment of modern society here–it's an eternal human misconception. Here is a slam on "Fratin" (originally written in Latin) from A.D. 63:

> Spoken Latin has picked up a passel of words considered too casual for written Latin, and the grammar people use when speaking has broken down. The masses barely use anything but the nominative and the accusative . . . it's gotten to the point that the student of Latin is writing in what is to them an artificial language, and it is an effort for him to recite in it decently.

Yet it is through just the kind of thing that seems so slovenly today that the language of tomorrow develops. Language change, to the extent that we can perceive it, appears to be decay. And sometimes it is, in the technical sense (erosion of sounds, dropping of endings). But in fact it concurrently entails building up (new sounds, grammaticalization of concrete words into new helping verbs, prefixes and suffixes) and plain old reshuffling *(nickname, I have seen her)*. No scholar has yet encountered a forlorn culture where the language simply "wore down" to the point that the people can no longer communicate beyond desperate barks (not even English, contrary to ever-popular belief). Language change is neither decay nor even evolution; rather, it is transformation–a term I have deliberately used in place of *evolution,* with its connotation of progress. In all languages, the five processes of language change described herein, interacting constantly, maintain a fundamental level of complexity suitable to human expression.

The Legs of the Whale: Language Change Leaves Footprints

Today's languages are Polaroid snapshots of ever-mutating transformations of the first language in six thousand different directions. As we would predict, evidence of preceding stages melts away only gradually. Just as whales retain vestigial leg bones underneath their skin as evidence of their status as one of many transformations of the original terrestrial Ur-mammal, words, sentence structures, and meanings in a language conceal evidence of preceding stages in the language's history. These retentions are often quite surprising to the modern speaker. Some of my favorite examples:

When I was little, one of the most dramatic insults girls in my neighborhood would come up with to hurl at one another in an argument was *hussy*, although I don't think any of us really knew what the word meant. Now we all do, but very few of us ever have occasion to know that the word began as *housewife*, or more properly *hūswīf*; sound change transformed this into *huzzif*, where its roots in *house* and *wife* became obscure and encouraged the meaning to start drifting. The usual assumption today is that the hussy is probably not a wife!

We say *Bye!* every day, but have you ever wondered what a *bye* was or how a *bye* could be *good*? Believe it or not, the expression began as *God be with you*. As late as just a few centuries ago, English speakers were still aware that *Goodbye* was a rebracketed version of this phrase, such that in 1659, for example, it was spelled *God b'wy*, while the playwright William Congreve wrote our *Bye!* as *B'w'y* in 1687. Thus *Bye-bye*, derived from shortening and then doubling *Goodbye*, is strictly speaking a truncated form of "Be with you, be with you!" (which actually is a rather richly interpretable and apt thing to say to someone upon leave taking if you think about it).

What's up with *goose* and *geese*? It's a remnant of an accident. Way back before English was even called English, the word for *goose* was *gos*, and you pluralized it with an *-i*, as *gosi*. Through time, speakers have a way of altering sounds in words to be more compatible with one another for ease of pronunciation: early Latin for *impossible* is *inpossibilis*, but it evolved into *impossibilis*, because *p* is pronounced with the lips, and if you think about it, the *m* sound is just a version of the *n* sound produced with the lips.

In reference to our *gosi*, pronounce *ee* and notice that it is closer to the front of the mouth than *oh*. While saying *gosi*, speakers would anticipate the frontness of the final *ee* sound by pulling the *oh* sound somewhat toward the front—*ee* pulled on *oh* like a magnet. You can produce the sound that resulted by shaping your mouth for "gose" and saying "guess" (this is the sound of the vowel in French *peur*). This meant that now "goose" was *gos* but "geese" was *gøsi*, where *ø* is our symbol for that bastard child of *oh* and *eh*.

Meanwhile, the grand old tendency in sound change to erode unaccented final vowels played its hand and took away the final *-i*. This left a situation where *ø* was the *only* thing distinguishing the singular *gos* from the plural *gøs*. Gradually the *ø* sound evolved into an *eh*, and then this gradually became an *ee* sound. Thus we have today's *goose* and *geese,* an accidental, cumulative result of a sequence of ordinary sound changes. No language ever makes perfect "sense" across the board, because all languages drag around "junk" leftover from the old days.

Just as *name* is written with a final *-e* because when the spelling was established the *-e* was still pronounced [NAH-muh], final *-e*'s in French represent a stage intermediate between Latin's case endings and the current situation in which the *e*'s are not pronounced at all. Latin had *parabulat*, and then one stage of "Fratin" had *parle* pronounced roughly "PAR-luh"; today the spelling is frozen by tradition, but the word is pronounced just "PARL." Yet the French preserve an echo of "Fratin" in this vein–but only in poetry and song. The ditty *Alouette/gentille alouette* . . . is pronounced "ah-loo-ET-uh/zho^ng-tee ah-loo-ET-uh" to fill out the lines of the melody, despite the fact that a French birdwatcher pointing out a lark shouts "ah-loo-ET!" not "Là-bas, Jacques! Une ah-loo-ET-*uh*!"[7]

Remember that *ilk* for *each* in the Lord's Prayer excerpt from Middle English? It had two fates. On the one hand, it evolved into our *each.* On the other, it was the source of the final *-y* in today's *every.* There was no single-word ancestor to *every* in Old English along the lines of something like "ælferiglic." *Every* began as a two-word expression *ever each, æfre ælc.* By Middle English, the two-word expression had been rebracketed as *everich*, and it was a short step from there to our little *every.*

There's *warmth* from *warm* and *growth* from *grow*–why not *slowth* from *slow*? Well, actually there is, but sound change has turned *slowth* into *sloth*, whereas semantic change has narrowed its connotation to a moral rather than aerodynamic one, thus obscuring the connection to *slow* for anyone but the compulsive.

It's like Charlie Brown. Charlie Brown is a bald child. Did you ever think about that? Charlie Brown is an eight-year-old who has virtually no hair on his head! This is something that we, after fifty years during which *Peanuts* has been an American institution, just

7. Somehow you don't imagine that the French are much into birdwatching, actually.

accept. Yet, of course, we can all presumably accept that, in life as we know it, the hairless eight-year-old is vanishingly rare.

The reason Charlie Brown has no hair is that, until roughly the mid-1960s, in comics and cartoons baldness was a kind of established signifier for dopiness—when someone drew a guy as bald, it *meant* something specific. Early comic-strip sensation Barney Google ("with the goo-goo-googly eyes") was a bald man. A Richard Deacon with hair would probably not have been hired to play office-butt-of-jokes Mel Cooley on *The Dick Van Dyke Show.* But today, on *Frasier,* Bulldog's baldness is entirely incidental to the character and is never milked for laughs.

So when *Peanuts* began in 1950, it was quite natural, part of the American pop iconographic language, for neighborhood reject Charlie Brown to be bald—this had a clear "meaning" in 1950: this kid is dopey. As time went by, however, the baldness ceased to have any "meaning." We no longer "read" baldness as an indication of laughability in men: if anything, it is now considered sexy—witness Bruce Willis in *Pulp Fiction.* Charlie Brown's baldness—which certainly does not evoke in us speculations about his bedroom prowess—evolved from a character marker into just part of the scenery as the strip evolved with the century, carried along as a "that's just the way it is" feature like *geese* and *sloth.* Babies were born (Linus and Sally started out as toddlers in the strip), fashions changed (some characters eventually wear leggings and sweatsuits), early characters faded into the background (by the end, vintage characters like Shermy and Violet barely existed except in crowd scenes), and so on. By the end of the run in 1999, *Peanuts* was only faintly recognizable as the strip that had debuted in 1950. But, throughout, Charlie Brown remained bald, as a reflexive remnant of an earlier meaning now bleached out and lost. Languages are chock-full of Charlie Brown heads.[8]

It's a Wonderful Language: Language Change Can Proceed in Myriad Directions

I mentioned in the Introduction that the fit between language evolution and biological evolution is not perfect. The evolution of animals and plants is partly constrained by the niches they inhabit: a

8. Never again will that sequence of words be used in the English language.

creature that takes advantage of resources available in the air, such as flying insects, may well evolve wings in one way or another. Language evolution, on the other hand, is largely a matter of chance, like the eternal transformations of that clump of lava in a lava lamp. Each of the changes in English and French that I have described were one of several possible pathways that could have been taken.

Because of this, as often as not, a language evolves not into just one language, but several. This happens when speakers of a language move to different locations: in each place, different change pathways happen to be taken, each setting the scene for new changes in their turn, such that, through time, the various transformations of the original language are no longer mutually intelligible.

Thus Latin evolved not only into French, but in other places into other languages, collectively termed the Romance languages: Spanish, Italian, Portuguese, and Romanian are the ones with the most speakers, while Occitan of southern France, Catalan sharing space with Spanish in Spain (and poking a bit into France), and the various obscure mountain varieties called Rhaeto-Romance in Italy and Switzerland (which include the "other" of Switzerland's four languages, Romansch) also survive. The French, Spanish, and Italian *Hamlet* passages are clearly in quite distinct languages, and yet all three evolved from Latin through the processes of change we have seen.

To see how such different languages could arise from one source, let's go back to our Latin verb "to have," *habēre*, pronounced "ha-BAY-reh." Each Romance language has transformed this word somewhat, but each in its own, unpredictable way. Spanish today has *haber*. The *h* in the spelling has not been pronounced for centuries, dropped Eliza Doolittle–style centuries ago, and Spanish let go of the final *-e* just as we let the one in *name* go, the result being "ah-BEHR." But in Italian, the wind blew in a different direction, creating *avere* [ah-VAY-reh]. It let the *h* go, too, but kept the final *-e*, instead "softening" the *b* to a *v*. Meanwhile, Portuguese went the Spanish route and then softened the *b* as well: it has *haver* [ah-VEHR]. Romanian really lived life to its fullest and stripped the word down to *avea* [ah-VAY-ah]. French has transformed *habēre* the furthest: out with the *h* and the final *-e*, soften the *b* to a *v*, and then transform the first *e* into "wah," and, voilà, *avoir*: "ha-BAY-reh" becomes "ah-VWAR."

The languages also differed in how they treated the word in regard to semantics and sentence structure. In Spanish and Portuguese, the word that evolved from *habēre* no longer means "to possess" in the literal sense. Instead, the Latin verb for "to hold," *tenēre,* broadened to mean "to possess," since holding something as often as not implies possessing it. Thus "to have" is *tener* in Spanish and *ter* in Portuguese. We saw how French yoked *habēre* into its *have*-perfect construction; the other Romance languages did this, too.

In fact, any language is most likely one of a litter of pups, and Latin itself was no exception. Latin began as one of several transformations of a language of which no records remain, but which we can deductively reconstruct from similarities between it and several dozen other languages in Europe, Iran, and India, collectively termed the Indo-European family. These languages are too similar not to have developed from one ancestral language, just as it is clear, despite their differences, that pigeons, penguins, and ostriches are all related by a single original ancestor. In the first edition of *The Origin of Species,* Darwin presented a prescient idea of what whales' ancestor would have been like, through deduction from the anatomy and behavior of whales and other mammals. His reconstruction of a queer semiaquatic bear with a big mouth has since been confirmed by fossil finds.

In the same vein, linguists have been able to reconstruct an approximation of what Proto-Indo-European would have looked like. When we see that awesomely distinct tongues like the Romance languages, German, Swedish, Russian, Polish, Irish, Greek, Persian, Albanian, Hindi, Bengali, Lithuanian, and dozens of others, developed from one parent, we see that from *habēre* to *avoir* is only the tip of the iceberg when it comes to the fundamental changeability of human speech.

Here is *sister-in-law* in seven Indo-European languages:

Sanskrit:	*snuṣā́*	Greek:	*nuós*
Old English:	*snoru*	Armenian:	*nu*
Russian:	*sňokhá*	Albanian:	*nuse*
Latin:	*nurus*		

(Actually, good old semantic change has transformed the meaning to *bride* in the last two.) How do we figure out what the original word was? For example, was the first vowel *o* or *u*? Well, only two have *o*, and it is more likely that a *couple* of languages changed *u* to *o* (*oo* to *oh* is a pretty short jump–just put your lips together . . . and then don't squeeze so hard) than that *several* changed *o* to *u*, so *u* is a better choice.

Next, did the protoword begin with *sn* or *n*? Here, "majority rules" is not as appropriate; it is much more likely that an original *s* wore off in several languages than that *s*, and always *s*, mysteriously just appeared out of nowhere in India, England, and Russia.

Now, what about that consonant nearest to the end of the word? Three words have *s*: Sanskrit, Greek, and Albanian; if only another one did, then we could go by "majority rules" again. As it happens, we can nudge the data in that direction. In certain positions, *s* in Latin had a way of turning into *r*, meaning that the *r* in *nurus* can plausibly have begun as *s*. (To explain why this was the case in Latin would be too lengthy a digression, but as a teaser, Latin's *honor*, the source of the same word in English, began as *honos*.) Ecce majority rules–now we have "snus" as our first four sounds in the protoword.

Finally, what was the ending? Since we're dealing with a female, we assume the original ending must have been the *-a* that we see in Sanskrit and Russian. But we can't ignore that Latin and Greek have *-us* and *-os*, which are masculine endings! To push the envelope, Armenian has this same kink, only in private: *nu* when endings are attached becomes *nuo*, the *o* being an echo of a masculine ending. "One never knows, do one?" as Fats Waller used to say, but at least in my personal experience sisters-in-law are generally women. Yet by "majority rules," the original Proto-Indo-European word for *sister-in-law* appears to have been masculine! Really, as counterintuitive as this seems, even general logic requires us to accept it: it is much more plausible that a couple of languages changed a masculine ending to a feminine one to bring the word into line with common experience than that cultures in seven different locations all changed the ending to a masculine one, with all allowances made for diversity among cultures.

How many licks does it take to get to the center of a Tootsie Roll pop? "The world will never know," as they used to say in the old Saturday morning commercial, nor will it know why the Proto-Indo-Europeans referred to their sisters-in-law with a masculine ending (the subject would furnish a rich basis for a nice indie film by Jane Campion). But apparently they did, and the most plausible original source for what today are *snuṣā́, snoru, snokhá, nurus, nuós, nu,* and *nuse* is *snusos*.

In different places, *snusos* underwent various combinations of myriad alternative changes, the result being words often barely recognizable as related (to someone raised on *nu*, *snokhá* would not exactly feel like something to grab onto when learning Russian). Latin developed through the same transformation of Proto-Indo-European that it would itself later undergo to become French and the other Romance languages. Romance future-tense endings like Italian *amerò* "I will love" developed by the squishing together of Latin's *amare* and *habeō* "to have." Latin had had its own future-tense endings: "I will love" was *amabo*. These were in turn eroded leftovers of what began as whole words in Proto-Indo-European: *amabo* began as *am* "love" and a word pronounced roughly "bwo" "I am"; this "bwo" hooked on to *am* and gradually lost its mojo and became a mere appendage.

But then, even Proto-Indo-European showed signs of being just a day in the life of an ever-changing system. The case endings that bedevil the learner in Latin (and Sanskrit, Russian, and so forth) did not emerge amid the Ancient Romans: Proto-Indo-European itself was bristling with case endings on nouns and endings for person and number on verbs and the like. From tracing the development of endings of this kind in other languages over time, we know that they almost always develop from what begin as free words, like the Romance future endings beginning as forms of *habēre*. Therefore, the endings in Proto-Indo-European, since they, too, must have begun as separate words, are signs that this language, too, was just one more of thousands of end products of millennia of change from the Ur-language.

And now we can pull the lens back even farther and view the world's six thousand languages as accumulations of endless transformations of the single African progenitor of all of them. As the first language evolved by the time it hit Nigeria, the result was languages like Yoruba, in which there are no consonant clusters and no case endings but words have different meanings depending on whether you pronounce them with a high or a low tone: *fi* with a low tone means "to swing," but *fi* with a high tone means "to dry." Speak Yoruba in a monotone and the ambiguity leaves you somewhere between a Steely Dan lyric and insulting the listener's relatives. In Russia, people speak a variation on the first language in which the erosion of many vowels has made consonant clusters like *vzgly* ordinary (*vzgljad,* pronounced "vzglyat," is "look, glance"; the word can even be romantic–"Just one 'vzglyat'/ That's all it took . . ."), and nouns come in three genders, each with sets of six different case endings that often differ from gender to gender.

Up in icy northern Canada and Alaska, the original language has evolved such that whole sentences can consist of single monster words. In Greenlandic Eskimo, "I should stop drinking" is *Imin-ngernaveersaartunngortussaavunga* (just as often, Eskimo sentences do have more than one word; it's just that the Eskimos' conception of how much you can pack into one word is more expansive than ours). Down in the middle of Australia, another variation on that African protolanguage, Jingulu, has shed all of its verbs but just three–*come, go,* and *do.* To express any action but those three (that is, almost every time you open your mouth–how often do you find yourself saying, "Oh, there's nothing quite like *doing*"?), you have to use an expression combining one of these verbs with a noun: you don't dive, you "go a dive"; you don't sleep, you "do a sleep." On top of this, the language has no word order whatsoever: what we would process as "word soup" is everyday speech to the Jingulu.

Language Is a Lava Lamp: Language Change Is Only Marginally a "Cultural" Phenomenon

The central role of chance in how language changes leads to another important point: language change is a culturally determined phenomenon only to a marginal extent. Culture appears central to the process within our life spans, because new words and expressions specifically indexed to cultural changes appear and disappear by the month. However, "culture changes and the language changes with it," while in itself true, is not the lesson I am hoping to impart, because culture only drives a sliver of the whole of the change that any language is always undergoing.

Certainly, languages have words or expressions expressing culturally embedded concepts for which there is no word in other languages. German's *gemütlich* is nominally "comfortable" but conveys a certain "Let it be Lowenbrau"/"What more could you ask for?" comfort you feel with good friends for which no English word quite suffices. *Comfy* and *cozy* are too apple-cheeked, *mellow* implies a low key that *Gemütlichkeit* does not necessarily require (it can just as well apply to jolly boisterous conversation), *comfortable* makes you think of a seat, and so on. Examples like this go on and on. But *gemütlich, macho,* and *ennui* are scattered examples; rarely can one persuasively index culture to the way a language expresses core concepts like "dog" or "drink," even if the treatment of these concepts differs slightly from ours.[9]

Thus, although the precise meanings of seemingly "basic" words can vary from language to language, these cases rarely have anything to do with culture, instead stemming from the inherent randomness of general language change. In French, we are often taught that *sortir* means "to leave." But more precisely it means what we mean by "go out"–"When do you want to leave?" is *Quand veux-tu partir?* and *Quand veux-tu sortir?* is "When do you want to go out?" Furthermore, the French can use *sortir* in an extended sense to mean "to pull out," as in "to make something go or come out": *Sors moi d'ici!* "Get me out of here!" Thus French's *sortir* occupies a semantic space that overlaps with, but does not coincide with, our *leave.* But there is nothing inherent to the French as a culture that led to their using the same word to describe both going out on a date and being pulled out of a hole. Nor is there anything inherently "French" about "taking" a beverage (*je prends une bière* "I take a beer" for "I'm drinking a beer") instead of drinking it or inherently English about "taking" a nap, whereas the French "make" one (*faire un petit somme*).

Even where culture does have an effect on core words, it is, cross-linguistically speaking, by no means a regular association seen regularly in all human groups with same relevant cultural parameter. Japanese has different forms of many verbs depending on the power relations between people in the exchange. To discuss eating with a peer, one uses the verb *taberu*; however, to talk about eating in a conversation with a superior personnage–or

9. Many readers will at this point be thinking about Eskimos and their words for *snow.* One just *wants* such a neat idea to be true, but sorry, folks. Let me do my part here to help dispel this myth: the idea that Eskimos have a plethora of words for *snow* is a mistake perpetuated by increasingly distorted generations of citations in the past century. The original source that this misconception stems from cites only four words for *snow,* and even this must be seen within light of the fact that we have *snow, slush, sleet,* and *blizzard,* despite English speakers having traditionally had neither any particular fascination with nor any significant cultural rootedness in snow. See Geoffrey K. Pullum's *The Great Eskimo Vocabulary Hoax and Other Irreverent Essays on the Study of Language* (Chicago: University of Chicago Press, 1991) for a tart, funny, and authoritative deconstruction of this myth.

even to talk to someone else about having eaten in the presence of said personnage–the verb is *itadaku*, whereas in referring to that higher personage's eating (whether that person is present or not) one uses a different verb, *meshiagaru*. Obviously, these verb forms are manifestations of the hierarchical nature of traditional Japanese culture.

Javanese (note the *v*; now we're in Java) really takes this ball and runs with it: a sentence as addressed to a member of the elite can differ so much from how one would say the same thing to a peasant that the sentences might as well be in different languages:

High:	Menapa	pandjenengan	badé	dahar	sekul	kalijan	kaspé samenika?
Low:	Apa	kowé	arep	mangan	sega	lan	kaspé saiki?
	Are	you	going	to eat	rice	and	cassava now?

In the grand scheme of things, however, these phenomena must be seen in perspective: there are plenty of highly hierarchical cultures whose languages do *not* correspond to social divisions in their vocabulary to nearly the extent that Japanese or Javanese do. A good example is Hindi, spoken in India, a country notorious for its rigid caste divisions. Certainly, one does not speak to members of a higher caste in the same way as one speaks to members of a lower caste. But the differences do not approach the likes of our Javanese rice and cassava example, being more akin to things like "Would Your Highness like tea?" versus "Hey, you! Out of the parking lot!" in English.

More to the point, though, a language consists not only of isolated words but also sounds and sentence structures, and these are at all times changing along with the word meanings. Here, culture rarely plays any role in a process that is essentially self-driven.

For example, there are languages spoken in arid regions of southern Africa with "click" sounds (as heard in the film *The Gods Must Be Crazy* or as used by the Ethiopian, Marvin, on *South Park,* although there are actually no click languages in Ethiopia). These clicks are not decorations; rather they determine meaning just as *b* distinguishes *bag* from *sag*. In addition, there are many different clicks, symbolized by linguists with various signs fortuitously reminiscent of comic-strip characters' cursing. In Nama of Namibia, *hara* alone means "swallow," whereas *!hara* is "check out," *|hara* "dangle," and *ǂhara* "repulse." The *xh-* in Nelson Mandela's native language Xhosa actually stands for one of these click sounds. Because all languages are and have always been in a state of continual transformation, anything we see in a language today is the result of change. These clicks, then, are products of sound change, but that evolution cannot have had anything to do with "culture."[10] What, precisely, about life in a hunter-gatherer society in an arid environment would make it somehow natural, necessary, or advantageous to use different click sounds to distinguish basic meanings like "swallow" and "repulse"? Remember that the clicks are not used to call people out of sight or to be "expressive"–they are ordinary sounds, distinguishing ordinary words just as conventional consonants and vowels do. Linking the clicks to these peoples' cultures is especially difficult given that people living in similar environments elsewhere, such as Mongolia, have no clicks in their languages. Plenty of words and expressions in Nama reflect its speakers' specific culture, but the clicks just crept in as unconnected to anything in the Nama soul as *geese*-style plurals were to the English soul.

As another example, German is well known for (often) placing verbs at the end of its sentences, but this is actually commonplace worldwide: Japanese, Hindi, Mongolian, Amharic in Ethiopia, Mandinka in Senegal, Nama in Namibia, and thousands of other languages do this, too. It would be hard to identify what "cultural" factor all these peoples could have in common from Tokyo to the Gobi Desert to Addis Ababa to Berlin that would explain this similarity. Or, to bring it closer to home, what was it about German history that would lead Germans–but not the British, Portuguese, Serbs, or Greeks–to start holding off on spitting out their verbs until they were finished with their sentences? ("Boy, Jürgen, we've sure a lot of beer d-d-drunk!")

Thus language change is hardly independent from culture, but it is in no sense collapsible into a larger category of cultural change. Language overlaps with culture but is not subsumed by it. More to the point, *most* aspects of any human speech variety's sounds, sentence structures, and word meanings are determined not by culture

10. However, it is not known just what sound-change process led to these clicks, and it has been suggested that the clicks were present in the *first* language and gradually disappeared in all of its offshoots except the few that have them today. But we still cannot say that the clicks disappeared elsewhere because they were incompatible with all cultures but a few in sub-Saharan Africa. Why couldn't a Kazakh click?

but by the cumulative effect of countless millennia of transformation proceeding through structured chaos. Although Japanese developed its hierarchical vocabulary alternatives because of the culture, in the meantime myriad sound changes, extensions, grammaticalizations, reanalyses, and random driftings in word meanings were taking place. The results of these processes constitute about 98.5 percent of the task in learning the language and had no more to do with Japanese culture than the rise of *Ich habe ein Bier getrunken* did to the German "soul." This is, and has always been, the case in all of the world's languages.

We cannot know what the words or structure of the first human language were, but we do know that one must have existed. Its outlines come gradually to us out of the mists of time. Our first indications are some possible handfuls of words deduced backward by comparing Proto-Indo-European and some other reconstructed family protolanguages. Our first *concrete* records of language come with the first written materials: Sumerian cuneiform inventories, Egyptian hieroglyphic narratives, Mayan inscriptions. By that time, the Ur-tongue, ever transforming over generations in each of the offshoots of the founding band of *Homo sapiens sapiens,* had already propagated into thousands of variations, which would all have been utterly incomprehensible to the East Africans who had developed their original progenitor. Egyptian obelisks, the *Rig-Veda,* the April 22, 1877, issue of the *New York Herald-Tribune*–all of this writing is mere snapshots of yesterday's cloud formations. Human speech is structured variation, like Haydn's string quartets or the images in a kaleidoscope after each shake: within the bounds of anatomy, human cognition, and the exigencies of social harmony, the first language took on a dazzling and infinite variety of permutations.

Now that we've come this far, would it beg the reader's forbearance if I revealed that, in the true sense, there is not even really such a thing as "a language" at all? It's the nature of language change that makes the concept of "a language" logically impossible, and your having read this chapter allows me to share the reasons for this impossibility in the next one and, in the process, fill in our story of what happened to the first language.

2

The Six Thousand Languages Develop into Clusters of Sublanguages

The first time I went to Germany, I was pleased to find that my German was good enough to order meals, get a room at the hostel, and even understand enough to not "stick out" too much in German-speaking social situations. However, my joy was short-lived. I went to Konstanz, a town in the south of Germany, to stay with some German college students I had met in the United States. I soon found that, as soon as the beer started flowing and the conviviality level went up, I quickly lost any ability to understand a word anybody said. I will never forget the evening I spent at a local pub with them, as joke after joke in what might as well have been Navajo to me elicited rafter-raising howls of laughter and calls for more beer. Yet to my knowledge not a word of Navajo was uttered that night in Konstanz: all of these people were speaking German–or at least what they called German.

What I had run up against is a fact about languages that modern American life tends to relegate to the margins of our consciousness. Atoms are not the irreducible entities that scientists once supposed; instead, atoms are complexes of subatomic particles. In the same way, viewed up close, most "languages" are actually bundles of variations on a general theme, dialects.

By this, I do not mean that there is "a language" that is surrounded by variations called "dialects." As will end up being a kind of mantra for this chapter, "dialects is all there is." One of a language's dialects is considered "the standard," but this anointment is a mere geopolitical or cultural accident. Standard German for

"They're scuttling our ship! We're going under!" is *Sie machen unser Schiff kaputt! Wir gehen unter!* In the dialect that frustrated me, Schwäbisch, it would be *Dia machat onser Schiffle he! Mir gangat onter!* To the German, real-world sociological associations make the second sentence leap out as "other," "quaint," "rustic," and perhaps even "not 'real' German." To a foreigner familiar with standard German, it looks just plain weird–a kind of twisted rendition of what we were taught as "German" in textbooks. But a Martian, presented with both sentences, would find no way of designating one as "the real one" and the other one as "a variation"; they would just look like two similar systems, just as a Burmese and a Siamese cat are to us different but equal versions of the same basic entity.

And in fact there *is* no "default" cat; there are only types of cat. Language change parallels biological evolution not only in creating different "languages" equivalent to species, but in that most languages consist of an array of dialects equivalent to subspecies. As such, not only did the first language evolve into six thousand different ones, but most of these in turn evolved into what, taken together, is an untold number of subvariations on those languages.

How Do Dialects Arise?

Dialects follow naturally from the inherently nondiscrete nature of language change. Latin developed into several distinct languages when populations of its speakers were dispersed throughout Europe. As we saw, however, the new languages appeared not abruptly, but by a gradual process in which there was no inherent dividing line between "Latin" and "new language." For centuries, the Latin of the region now known as France was a variety halfway between Latin and French, which I have facetiously called "Fratin." Similarly, in Spain there was once a "Spatin," in Italy a "Latalian," etc. In other words, had we toured Europe in roughly A.D. 1, we would have found varieties of Latin not so distinct as to strike us as "separate languages," but distinct from one another nevertheless. At the time, they were still, in our terminology, new *dialects* of Latin.

We have concentrated on language change as it operated in the past to lead to today's languages. Yet because human speech is inherently mutable, it follows that today's languages are slowly undergoing the same transformative process that Latin did. In this light, a question arises about, for example, English.

The English They Don't Want You to Know About

English is one of several languages that evolved from an unwritten ancestor linguists call Proto-Germanic; other Germanic languages include Dutch, Swedish, Danish, Yiddish, and (three guesses!) German. The different languages resulted from Proto-Germanic speakers settling in different locations, England being where English developed.

Okay, good–but obviously England is not a giant open field where all of its speakers interact with one another on a daily basis in a grand, mad, Woodstockian splash of teeming humanity. On the contrary, the Nordic invaders who conquered and took over the island in A.D. 449 quickly spread their language in all directions, and in subsequent centuries English came to be used in hundreds of separate regions, by people most of whom rarely ventured far from their villages.

We would expect, then, if Latin became several languages once it was spread among several separate populations, that English itself would have begun developing into different languages, just as Darwin's Galápagos finches began developing into distinct varieties once let loose on those islands–and this is exactly what happened. In the England of our moment, the process has not gone far enough to lead to separate languages. Instead, just as there was a time when today's Romance languages were still dialects of Latin, today's British English varieties are dialects of English, recognizable as "the same language" but quite distinct nevertheless.

It is important not to think of the regional varieties as having evolved *from* the Standard English we know today; the invaders of the continent did not arrive speaking like George Washington, or even Shakespeare or Chaucer. They arrived speaking Old English, and today's standard evolved *alongside* the ones that were eventually relegated to "regional" status. Thus the common source of all of today's English dialects is that queer-looking tongue in which *Beowulf* is written, a tongue that is no longer alive. Old English developed in different directions in each region, and thus each dialect developed its own sound changes, extensions, grammaticalizations, rebracketings, and semantic changes.

One gets hints of this on the British sitcom *Are You Being Served?* Ladies' Intimate Apparel sales assistant Miss Brahms is a Cockney and at first threw me a couple of times with sentences like *An' it's expensive an' all: an' all* has evolved semantically in Cockney to mean *too*. Ending a spectacular run in 1984, most of the cast members reunited to tape two more seasons in 1992, this time assigned to run an upcountry farm. The locals pronounced Mr. Humphries' name "Mister 'Oomphries"–just as *Frasier*'s Daphne would say it–because evolution of the *uh* sound here differed from that in other parts of England, and they said *sommet* for *something*, because the Old English source of *something* evolved through different sound changes in this area.[1]

Yet this kind of thing is just the tip of the iceberg. Here, for example, are some sentences in the English spoken until about a century ago in rural areas of the southwestern region of Cornwall:

Aw baint gwine for tell ee.

Th' Queeryans do s'poase the boanses ded b'long to a helk.

Ded um diggy ar no?

Billee, 'ome, d' b'long gwine long weth 'e's sister.

At first glance, the first sentence does not appear too opaque: "I'm not going to tell him," right? But no–it means "*He* isn't going to tell *you*"!: *aw* evolved from Old English's *hē* (pronounced "HAY") through different sound changes from those of Standard's *he*, and *ee* is an evolution of the initial *y* in *you.* And notice *for* instead of *to. Baint* is a rebracketing of *be* and *not* into a single form. Its ultimate source is extension: we tend not to notice that English uses no fewer than three different roots to inflect *to be*: *am/are/is, be/been*, and *was/were* derive from what used to be separate verbs, all of which inflected on their own in all six person/number combinations in all tenses. Standard English puts up with the odd division of labor among the three roots–no dialect, or language, irons out any but a fraction of the things that could be. But Cornwall English, like other

1. Amazingly, the revival, called *Grace and Favour* in England but *Are You Being Served Again* in the United States, was every bit as good as the original series, and it is to be hoped that the cast blesses us with at least one more go-round.

Southern dialects, extended *be* into negative constructions instead of only allowing *am/are/is* into that area; thus *I be not*, later *I baint.*

The second sentence is "The Antiquarians suppose the bones belonged to an elk." The *do* and the *did* are not meant for emphasis. The *do* was used quite neutrally, having been *grammaticalized* otherwise than in Standard English. Standard English uses *do* this way only in negative sentences: *I don't know*; in Cornwall you could also use it in affirmative sentences–what began as a full verb *do* evolved into a faceless little piece of grammar. The *ded* for *did* is a grammaticalized marker of past, just as our *-ed* marker is (it has been guessed that even *-ed* evolved from what began as *did*). *Helk* represents the extension, virus style, of initial *h* onto words that at first did not have it. In this as in many regional English dialects, sound change tended to erode *h* from the beginning of words (*'Ello, mate!*). Yet because changes go through a phase where the old and the new form exist side by side, speakers who tended to let the *h* go were aware that there was both an *h*-less and an "*h*-full" version of the word: *hello/'ello, horse/'orse.* Notice that dropping the *h* leaves the word with a vowel up front: for many speakers, a feeling set in that *any* vowel-initial word might have an *h*-full alternate, and the result was words like *helk.*

Ded um diggy ar no? is "Did he used to dig or not?" The *diggy* is not just a "cute" way of saying *dig* in the vein of "diggy-poo"; the sentence does not, for example, refer to a child. In Cornwall English, the suffix *-y* is a piece of grammar, used specifically to convey that an action is repeated. The fourth sentence is particularly confusing for us. The *b'long* does not literally refer to possession but is another marker of habituality, where the full meaning of *belong* has grammaticalized, and thus the sentence means "Billy, at home, usually goes with his sister."

This, then, is what happened to English as it mutated down in the southwest, as at the same time farther northeastward it was mutating into the standard variety we are familiar with. Meanwhile, up in the Midlands in Nottinghamshire, in local speech until not long ago one might hear *Tha mun come one naight ter th' cottage, afore tha goos; sholl ter?* Some readers might recognize this as one of the sentiments expressed by the gardener in *Lady Chatterley's Lover;* whereas Lady Chatterley at one point snaps, "Why can't you speak normal English?" the gardener's speech is simply the direction into which English had evolved in this region in contrast with her standard-speaking one. In Farnworth, north of Manchester, two forms of both *yes* and *no* have evolved. *Yes* is *aye* under usual circumstances, but if one is contradicting a negative statement, then the form is *yigh* (pronounced like *aye* but with an initial *y*): A: *I can't find the scissors.* B: *Yigh, they're here. No* means *no* usually, but you contradict a statement someone just made with *nay*: *Nay, by gum! I'm not having that!* (They really do say "by gum.") In French, *si* is used similarly: A: *Tu ne l'aimes pas.* B: *Si, je l'aime!* (*You don't love him. Yes, I do!*), as is German *doch.* There's no reason why English shouldn't have this; Standard English just happens not to have taken this route. Farnworth English, however, did. All non-Standard English dialects use "double negation" (*I ain't got none*), but Farnworth English also allows particularly spectacular negation Dagwood sandwiches: *I am not never going to do nowt no more for thee.* (*Nowt* is *nothing.*)

One can think of dialects as different recordings of a pop song. "I Say a Little Prayer" was first recorded in a perfect three minutes by Dionne Warwick. Later, Aretha Franklin did a version, an equally perfect, but different, three minutes. One day, Mariah Carey, in her quest for true diva-hood, will most likely record the song, and that will be another fine three minutes. Even Luciano Pavarotti might give it a go some day. There is no "blueprint" "I Say a Little Prayer" in the way that there is a "blueprint" "Sempre libera" from Verdi's *La Traviata,* solemnly imprinted with melody and accompaniment precisely specified. There is a sheet-music version of "I Say a Little Prayer" for piano, but it's just an anemic little toss-off by some anonymous house arranger at the publishing company, hardly as colorful as Burt Bacharach's orchestration for the Warwick recording. But then Bacharach did not mean this first recording as "The One," and put together different orchestrations of the song for later recordings. And then the sheet music is just designed to accompany a singer anyway–neither Bacharach nor anyone else would consider Aretha Franklin's rather free, often improvised approach to the melody a "violation of the composer's original intentions."

Properly, there *is* no "Ur-text" "I Say a Little Prayer"; it exists *only* as various interpretations of the basic outline. Standard German and the Schwäbisch dialect I ran up against or Standard English, Cornwall English, and Cockney English exist in the same relationship as the various renditions of that song. Moreover, there is no "broken down" dialect that stands in relation to the others as the party revelers' drunken, out-of-tune rendition of "I Say a Little Prayer" in the movie *My Best Friend's Wedding* does to the Warwick and Franklin recordings. Each dialect is just a different roll of the language-mutation dice.

Along these lines, as often as not, a language comes into existence split into different dialects from the very beginning, there never having existed any single original variety. Romans did not settle Gaul in a single clump, but spread out through the area, passing their language on to separate populations. There was enough intermigration and travel that the new Latins that developed in the northern area maintained a fundamental kinship with one another (whereas in the south things went so much their own way that a distinct language developed, now called Occitan); nevertheless, what became "French" differed significantly from region to region.

We saw how, in what I have called "French," the *k* sound often became the *sh* sound, such that Latin *canem* became French *chien*. This did not happen in all French dialects, though; it did in what is now the standard one, but in the northwestern regions of Normandy and Picardy, *k* stayed *k*. A piece of coal in Latin was *carbo*; in Standard French it is now *charbon* [shar-BAW^{ng}], but in Normandy and Picardy it stayed *carbon*. At first, French dialects had *ei* where the standard now has *oi* (pronounced "WAH"): the midpoint between Latin *habēre* and Standard French *avoir* was an earlier *aveir*. Because language change is a chance affair, in the Norman dialect the change to *oi* never happened to transpire. That is why, in my fourth-favorite city in the world, there is a Mount Royal (*royal* being the Standard French form), but the city it is in, settled by people most of whom came from northwestern France rather than Paris, is called *Montréal*, *réal* being *royal* in Norman French. Over in the east in Lorraine, a special past tense has grammaticalized through the evolution of *or*, which means roughly "now," into an ending: *j'étozor* means "I was" in situations where something else is about to break in on the proceedings. One of the oddest things about French for English speakers is learning that there is a special past tense used only in writing: one first learns that *he spoke* is *il a parlé*, but then finds *il parla* in novels and stuffy writings. In western and eastern dialects, however, these forms were commonly used in speech for centuries after they fell by the wayside in the standard.

Similar facts could be trotted out about most languages on earth spoken by groups larger than a village or two. Even Fijian, spoken on a complex of islands by just seven hundred thousand people, has more than one dialect. If geography or culture ensures that subsets of a group speaking a language interact and identify more with one another than with the larger set of all people who speak the language, the inevitable result is language change, as in cooking, art, music, and dance, developing in divergent directions. Outcome: dialects–Szechuan, Hunan; *Did he used to dig?, Ded um diggy?*

All Systems Normal–At Least for Now

One thing that follows simply and ineluctably from this is that, despite the almost irresistible pull of the sociologically based evaluations that attach to dialects, there is no such thing as human beings speaking "bad grammar." There are no dialects in any way analyzable as "decayed" versions of the standard or of anything else. Why would speech "decay" down in Cornwall but keep a stiff upper lip in the central Midlands? Why would Latin "crumble" in Picardy but for some reason just "evolve" around the Île-de-France?

It is almost sobering to realize that the social evaluations we place on how people talk are purely artificial constructs placed on speech varieties that neither a Martian nor often even a foreigner unfamiliar with our social terrain would arrive at on the basis of recordings of the speech alone. This observation goes down easy when we think about peasants in Picardy, of course–because we have not been steeped in the social evaluations particular to France. It is harder to truly wrap our heads around this here at home, though.

The Appalachian English that sounds so twangy, rustic, and full of "mistakes" (that is, ways in which it has mutated in directions other than Standard English has) would be just one more variety of English to our Martian–and we're talking Snuffy Smith and Li'l Abner here, not just a "country" accent. Black English, America's most controversial dialect, which even the most well intentioned of people often see as "bad grammar run wild," developed through the same processes of change as those of any other dialect and thus stands equal to any other in the qualitative sense.

I have never heard the common conception that nonstandard dialects are "bad grammar" put as eloquently as when an elderly black woman in the Mississippi Delta once said to me that, from what she saw, "Seems like most people speak pretty good English, but some people, it seems like they just talk!" To her, it naturally seemed as if the Southern Black English dialect spoken by many around her, especially in its "deeper" varieties, was "bad" English rather than alternate English. In this vein, Black English speakers are often accused of having "bad diction," but this is mainly a trait local to many black male teenagers' in-group identity and is common in dialects spoken by male teens in many societies (listen to them on the streets of Paris or Berlin); in any case, some whole language's sound systems are just "crisper" than others'. Brazilian Portuguese, for example, is the antithesis of German in this regard, often sounding as deliciously gushy as a ripe slice of mango (think Astrud Gilberto, and this is even more pronounced in running speech). Yet we would be hesitant to accuse the entire country of Brazil of having "slurred speech" in comparison with that of Spanish-speaking countries.

And notice that dialectal differences run wide and deep. "Dialect" is not meant here just as a stand-in for accent and scattered regional terms: things like *Ded um diggy?* go far beyond this into the heart of sentence structures and shapes of even basic vocabulary. In America, the difference between dialects is rather slight in relation to how widely dialects often diverge worldwide. American dialect differences are largely limited to peripheral vocabulary (*pop* instead of *soda*), minor sound differences (*greasy* versus *greazy,* the fact that for young Californians, *John, Sean, Ron,* and *Dawn* all rhyme, the second and fourth pronounced "Shahn" and "Don"), and a kind of generalized set of nonstandard speech forms such as *ain't*, double negations, use of *don't* with the third person singular (*Don't make no difference to me*), etc. Even Black English, though a tad more divergent from Standard English than this in sound and even in sentence structure, does not strain our sense of what "English" is.

But America is rather exceptionally bland in regard to dialect divergence, just as its flora and fauna pale in comparison with the riot of creatures in a tropical rain forest. In Konstanz, it wasn't only the slang that was throwing me; I could barely understand when someone asked me whether I wanted another beer (which I of course took, which dampened my comprehension even more). Nor was it just a matter of twisting my ear to an "accent": at one point in the evening, I seated myself behind one guy who was speaking in a clear, resonant voice at moderate speed and found that I could still barely make out a word he was saying. In Schwäbisch, on top of the slang, different sound changes and differing fates and redistributions of endings have rendered even basic words into shapes related to, but significantly distinct from, the ones that developed in the standard and have created different endings: standard's *wir gehen* [veer GAY-un] is *mir gangat* [meer GONG-uht], and so on.

We have never had the equivalent of this in America. We may have to adjust our ears to certain local dialects to an extent, but there is barely anywhere in the country that Standard English speakers could go where, after their hosts had had a few at the local pub, they would find themselves utterly at sea linguistically.[2] By the time America was founded, various aspects of modern societies that tend to retard dialects' mutation were long established. Widespread printing forces a decision about what "standard" speech will be and naturally has a way of enshrining that variety for future generations as "the language" because it lives immortally on paper. The spread of education, conducted in that standard variety, furthers this impression. Both factors pit the standard variety in competition with the local varieties and lead many speakers to lean their speech toward the former in deference to its association with prestige, which diminishes the extent to which the local variety changes in its own directions. Meanwhile, in the twentieth century, radio, television, and films imprinted the influence of Standard English even more deeply.

In Great Britain, English had more than a millennium to develop free from these artificial impediments to language change, the result being dialects such as Cornwall English, which would have baffled the foreigner trained in Standard English just as Schwäbisch did me and would have taken quite a bit of adjustment even for us. Meanwhile, though, the entire timeline of American English has taken place within the constraints of societal trends that have the by-product of retarding language change and thus the mutation rate of dialects.

Of course, printing, education, and the boob tube are as prevalent in Europe today as they are here. That is why I refer to Cornwall English in the past tense; throughout England, Standard English is strangling the old local varieties like kudzu, and thus, though I said that English dialects "have not" developed into sepa-

---

2. The exception would be places where either Gullah Creole or Hawaiian Creole English, generally called "pidgin," were spoken; on these and similar phenomena, wait until we get to Chapter 4.

rate languages with the implication that they eventually will, more properly, modern conditions are such that they never will. Increasingly, local differences are more a matter of local terms and accent than much else. It is now ever harder to find speakers under the age of 106 of regional French dialects as well: for example, no longer do you often hear *il parla* in small towns in the west or east of France.

The extent of this homogenization differs from country to country. Regional Italian dialects are still distinct enough that the student armed with Standard Italian can still have the same experience I had in Konstanz at a bar in Milan. The experience I had certainly did not suggest that Schwäbisch is exactly on the ropes, and recently in Leipzig, when electricians came to my office asking something in the local German dialect Sächsisch, the accent alone was so thick that I had to ask a German person down the hall to figure out what they were asking (I never did figure out what exactly they wanted to do). But even in Germany the dialects are diluting among younger generations. *Asterix* adventures have recently been published there in translations into more than a dozen "Mundarts" ("mouth-ways," or dialects) partly in an effort to help celebrate and preserve dialects widely felt to be in danger of imminent demise.

But the linguistically homogenizing tendencies of printing, education, and the communications revolution have set in only in the past few centuries, whereas human language has existed for about 150,000 years, as mentioned earlier. As such, for "languages" to consist of clusters of often highly divergent dialects has been the norm for human language for all but a final hiccup of its existence thus far. This is crucial in understanding that, because of the transformative nature of human speech, the concept of "language" is a mere terminological convenience. There is no intrinsically coherent entity that corresponds to our sense of what a "language" is. There is no heady, abstract, vaguely politicized philosophical argument behind this; it's really quite meat and potatoes.

Why Dialects Are All There Is

Martians Couldn't Tell: "Standard" Dialects Are Just Lucky

Because the standard variety is the vehicle of almost all writing and official discourse, it is natural for us to conceive of it as "the real deal" and nonstandard varieties as "other" and generally lesser, even if pleasantly quaint or familiar. This state of affairs also tends to foster the misconception that the standard dialect is developmentally primary as well: one can barely help operating on a background assumption that, at some time in the past, there was only the standard dialect but that, since then, nonstandard dialects have developed through the relaxation of the strictures of the standard.

But in fact standard dialects were generally only chosen for this role because they happened to be spoken by those who came into power as the nation coalesced into an administratively centralized political entity. What this means is that there is no logical conception of "language" as "proper" speech as distinguished from "quaint," "broken" varieties best kept down on the farm or over on the other side of the tracks.

The Right Place at the Right Time For example, today's Standard French began as just the dialect spoken in the area where Paris is today. It shared France with several other varieties, including those of Normandy, Picardy, and Lorraine, as well as the varieties of the south not even mutually intelligible with the northern ones and thus considered a different language, including Provençal, which was the vehicle of the love songs of the troubadors. Until the late 1700s, this linguistic heterogeneity was not considered a problem in France: in feudal times, the peasant's loyalty to the local lord, who probably spoke the same local dialect as the peasant did, was considered paramount and hardly depended on speaking the king's dialect.

But as the concept of nationalism began to arise, government officials began to be concerned that citizens of *La France* were not united by a common language. As one local official complained:

> The multiplicity of idioms could be used in the ninth century and during the overlong reign of feudalism. The former vassals gave up the satisfaction of changing their master for fear of having to change their speech. But today, when we all have the same law for master, today when we are no longer Rougeras, Burgundians etc., when we are all French, we must have only one common language, just as we all share a common heart.

The observation about peasants' "fear of having to change their speech" indicates how distinct many of these dialects were from one another: we're talking Schwäbisch versus Standard German, not "greazy" pork chops. And today's Standard French dialect was indeed a minority dialect in numerical terms: France was a dialectal smorgasbord. Abbé Grégoire, a Catholic priest and revolutionary, was alarmed that:

> France is home to perhaps 8 million subjects of which some can barely mumble a few malformed words or one or two disjointed sentences of our language: the rest know none at all. We know that in Lower Brittany, and beyond the Loire, in many places, the clergy is still obliged to preach in the local patois, for fear, if they spoke French, of not being understood.

And if there was a common language to be imposed, it was naturally to be the one used by those in power. Chauvinism was not the only root of this inclination: because the Paris dialect was the one that had been most written for centuries and used by those with the most power, it followed naturally to conceive of it as the "real" French.

From here on, this one of many dialects–now called "French"–was spread throughout France by education as well as an unfortunate dedication to eradicating the local dialects, dismissed as "patois," in the interest of national unity. Thus the dominance of today's Standard French in France resulted from an artificial perversion of an originally much more diverse scenario, rather as the rich fauna and flora of Madagascar have been significantly decimated by the roping of its residents into a global economy that drives them to clear-cut its forests.

In some cases, the standard dialect is even deliberately fabricated by picking and choosing from several local dialects. Today's Standard Finnish was deliberately codified first from southwestern dialects, with elements of eastern dialects interwoven in the 1800s with the popularity of the publication of the oral epic *Kalevala* in an eastern Finnish variety. The utter artificiality of this "standard" is shown by the fact that no Finn, of any educational or socioeconomic level, actually speaks this dialect casually. Newscasters, teachers, and sometimes politicians speak it as a deliberately neutral, "official" code, but there are no Finns who "offstage" speak a "vanilla" Finnish as, say, the characters in *Thirtysomething* or *That 70's Show*, who speak a faceless "generic" English. On the contrary, *any* Finn's regional origin is clear from his speech. Because no one speaks it except in formal contexts, the "standard" is considered not "the best Finnish" but a utilitarian strategy; Finnish dialects diverge significantly, necessitating an agreed-upon, even if arbitrary, common coin for official purposes.

Standard English developed through a kind of combination of the French and Finnish situations. Standard English incorporates elements from the Essex and Middlesex dialects that happened to be spoken in the London area. By the 1400s, London was the hub of manuscript copying and then printing. Because scribes and printers tended to come from the surrounding regions, features of dialects spoken in these areas made it onto the page more often than others, which in turn meant that people throughout England were more likely to see London dialect written than any other. Combine this with the cultural and commercial influence of London, and the result is a standard dialect perceived as the heart of "English," a star relegating the other varieties to character parts. The shift in attitude toward local speech is visible in the record. About 1490, England's first printer, William Caxton, depicted the differences between London and Kentish English as a matter of apples and oranges:

> In my days happened that certain merchants were in a ship in Thames . . . and went to land for to refresh them[selves]. And one of them named Sheffield, a mercer, came into an house and axed for meat. And specially he axed after eggs. And the good wife answered that she could speak no French. And the merchant was angry, for he also could speak no French, but would have had eggs and she understood him not. And then at last another said that he would have eyren, then the good wife said that she understood him well. Lo, what should a man in these days now write: eggs or eyren? Certainly it is hard to please every man by cause of diversity and change of language.[3]

---

3. I have modernized the spelling and pronunciation to make it easier on the eyes. The original spelling is the likes of: *And thenne at laste a nother sayd that he wolde haue eyren. . . .*

(Note those *axed*'s, which show how arbitrary our sense of "improper" English is; *axed* was accepted and ordinary even in written English at the time.)

But Caxton's equanimity was already on its way out; about 1400, a character in a play had casually dismissed northern dialects as "scharp, slitting, and frotynge and unschape," whose meaning comes through even without the modern equivalent "shrill, cutting, and grating and ill formed," and this kind of judgment was soon commonplace in England. Thus only chance determines that we have eggs Benedict rather than eyren Benedict and process *eyren* as vaguely "unschape."

Today's "Dialect" Is Tomorrow's "Language" Not only has one of many hitherto unranked dialects often been anointed the standard, but we even see dialects actively dismissed as "quaint vernaculars" at point A only to be enshrined as inherently noble vehicles of humans' loftiest thoughts at point B, with nothing but a decisive geopolitical shift at the root of the mysterious change in perception.

Dante wrote at a transitional period between point A and point B in Italy. Latin was still considered the appropriate vehicle of writing, officialdom, and educated discourse, but Italian had developed so far from Latin that its traditional classification as "village Latin" had come to strain natural perception. Dante, afflicted with that queer medieval southern European malady called courtly love, in 1293 dedicated a volume of poems to his adored Beatrice, who combined two traits unusual in a dedicatee of love poetry–namely, having never been touched by the author at any point in her life and being dead. Dante wrote this *La Vita Nuova* in Italian, but only as an extravagant gesture to a woman who understood Latin only with difficulty (not being alive probably made it even more of a chore). Dante usually wrote in Latin, and writing in Italian at all in this period was a nervy gesture. Yet his reason for writing *The Divine Comedy* in Italian in 1308 revealed a similar guiding sense that, at the end of the day, there was something fundamentally "kitchen sink" about Italian:

> From this it is evident why the present work is called a comedy. For if we consider the theme, in its beginning it is horrible and foul, because it is Hell; in its ending fortunate, desirable, and joyful,

> because it is Paradise; and if we consider the style of language, the style is lowly and humble, because it is the vulgar tongue, in which even housewives can converse.

Indeed, Dante's main grounds for championing Italian were practical; in his *De vulgari eloquentia* (note the Latin title), he urges that Italian be used in literature for the mundane reason that more people understand it than do Latin, which he meanwhile exults as nevertheless the "better" language. Yet though in the 1300s even one of the most masterful bards who ever blessed the Italian language essentially considered it a matter of "Comedy Tonight!", by the late 1700s, Lorenzo da Ponte, writing the lyrics and libretti of operas such as *The Marriage of Figaro* and *Don Giovanni,* would have been surprised to be told that he was writing in "housewives' Latin." By then, Italian was considered by its speakers and beyond as one of the world's loveliest, most singable and romantic languages. Only the gradual unification of Italy and its ascendance as a world power made the difference.

The Romanian-speaking area extends eastward into a little hump of land called Moldova, much of which for decades was incorporated within the Soviet Union. Moldovan is not just "close" to the Romanian dialects in Romania proper: it is very much one of them, not differing from the standard dialect any more than any Romanian nonstandard one does. The only remotely salient difference between Moldovan and Standard Romanian is that a polite form of the pronoun *he* in the standard is used more informally in Moldovan. Otherwise, most of the differences are minor differences in vocabulary no more dramatic than the ones between American and British English. Within the Romanian-speaking orbit, then, Moldovan had been a "quaint," "rustic" dialect. The Soviets, however, in a quest to discourage Moldovans from identifying with their Romance-speaking neighbors to the west, directly required Russian linguists to foster a conception of Moldovan as a "different language" from Romanian, exaggerating the import of the minor differences inevitable between dialects of any language. Many grammar books of "Moldovan" were little more than translations of Romanian-language Romanian grammars into Russian. Now independent, the Moldovans continue to encourage a perception of "Moldovan" as a distinct "language" from Romanian, in part because Romanians tend to dismiss their dialect as sounding uneducated. Hence the Moldovan "language," fully intelligible with Romanian right next door.

Don't tell the Scandinavians I said this, but "Swedish," "Norwegian," and "Danish" are all really one "language," "Scandinavian"– people speaking these "languages" can converse. Here is "He said he couldn't come" in all three:

Swedish:	Han sade att han inte kunde komma.
Danish:	Han sagde at han ikke kunne komme.
Norwegian:	Han sa at han ikke kunne komme.

These are even closer than Standard German and Schwäbisch or Standard Italian and Milanese. The Danes used to run what is now Sweden and Norway, and there was no such thing as "Swedish" until Sweden became independent in 1526. What is today "Norwegian" was just "the way they speak Danish in Norway" until Norway broke with Denmark in 1814 and gradually began explicitly working out a standard form of what was an array of nonstandard local dialects.

I once asked two Bulgarians what Macedonian sounded like to them, and they said in unison, "It's a dialect of Bulgarian!" "Macedonian" is indeed so close to Bulgarian that Bulgarians crossing the border need make even less adjustment than Swedes make in going to Denmark. Many Macedonians would find my Bulgarian friends' comment a little irritating, which stems from the fact that "Macedonian" is considered a separate "language" owing to its speakers' distinct political and cultural identity from Bulgarians, reinforced by their incorporation until recently into the Yugoslavian federation.

Although one learns Hindi in different courses from those in which one learns Urdu, the two are dialects of the same language. Hindi, the indigenous lingua franca of India, has taken on a lot of vocabulary from Sanskrit, its ancestor now enshrined in liturgical writings, and Urdu, spoken in Muslim-dominated Pakistan, has done the same from Islam's vehicle, Arabic. Yet this is little more of a barrier to basic communication than, again, that between American English and British English. The sense of separateness conditioned by the profound animosity between Hindu India and Muslim Pakistan extends to linguistic identity and encourages a sense of separateness between these two mutually intelligible varieties.

Today's "Language" Is Tomorrow's "Dialect" Conversely, there are also cases when a speech variety treated as a "language" at point A is suddenly a "dialect" in the history books at point B, again with mundane events rather than anything inherent to the speech itself having made the difference. Dante wasn't the only medieval European in the throes of courtly love: itinerant musicians in southern France made careers of composing songs to unattainable women of high rank, as the famous troubadors. They did so not in Parisian French but in the particular transformation of Latin that happened to have taken place in their region. Whereas dialects in areas surrounding Paris were similar enough to Parisian to be classifiable as "kinds of French," the dialects of the south were distinct enough to be processed as a different "language" altogether. The dialects of the north were called *langue d'oïl* "*oui* language," in reference to the word for *yes* in those dialects (not yet evolved to the modern *oui*), whereas the southern ones were correspondingly called *langue d'oc* (hence the region called Languedoc today). Note that in the Gallic consciousness of the period there was no inherent rank implied: the two "langues," or languages, were separate but equal. The dialect of the *langue d'oc* used by the troubadors was called Provençal, and from the 1100s to the 1300s was considered very much a "language," written as well as sung.

Troubadors are *trouvères* [troo-VAIR] in Standard French: the word is the standard dialect's descendant of a Latin root *trobare* "to compose," whereas *troubador* is how the same root came out in Provençal. This difference is an indication that Provençal was no "kind of French" by any standard; it was very much a horse of a different color. Similarly, in the modern *langue d'oc* descendant (Occitan), *uèch* is eight where French has *huit*, and so on.

But as the center of power concentrated increasingly on Paris, southern France was deliberately yoked politically and administratively into the "French" orbit, complete with transplanted French-speaking officials. This nationalist tide turned against the nations-within-nations that could foster alternate standard dialects, and Provençal and its *langue d'oc* kindred dialects were effectively banished from writing and official contexts. The scene was set for these dialects to be classified as "lesser" rather than "different," concretized by the post-Revolutionary language homogenization policy. By the 1700s, the once-prestigious "language" Provençal was a complex of rural dialects considered mere "patois." The general sense was that these dialects were a "kind of French" when, as we see, though there are no dividing lines to be drawn between "dialect of A" and "language B," Provençal was obviously different enough from Parisian French and the other northern dialects to fall on the "language B" side of the line. Thus the suppression of Provençal was less the silencing of one variety of "French"–sad enough in itself–but of a separate Romance language entirely. We'd rather that a particular subspecies of brown sparrow not become extinct, but the loss is perhaps even greater if we lose *all* subspecies of an entirely separate species of bird, such as pigeons.

Then, in the late 1800s, the poet Frédéric Mistral began a movement to revive these dialects, under the heading of a different name, Occitan. As a result, Occitan is now officially treated as "a language" again, complete with self-teaching materials, novels, and poetry. This cycle eloquently demonstrates that, in the end, dialects are all there is: the "language" part is just politics!

The Ukrainian "language" is a similar story. In Russia, one does not find Russian dialects as distinct from the standard as in Germany and Italy. The main reason for this is that history happens to have fenced off the regions where such "dialects" are spoken as separate cultural and political units, the "dialects" thus officialized as these units' separate "languages." Before the Ukraine was transformed into one of the old S.S.R.s, for instance, it was simply a region of Russia. As a matter of fact, when its city Kiev was considered one of the leading urban centers of Russia early in the last millennium, the dialects spoken there were the closest thing to any conception of "the best" spoken *Russian*, not "Ukrainian," because they were the ones most frequently written.

Starting in the 1300s, when Moscow became Russia's center of government, what is today the Ukrainian "language" became considered a peasant variety of Russian. The difference between Russian and Ukrainian is about the same as that between Standard German and Schwäbisch, often less. Russian for *get married* (when it's a woman doing so) is *vyjti zamuž*; the Ukrainian is *viiti zamiž*; the woman herself is *žena* in Russian, while *žona* is one word for woman in Ukrainian.

In college I had a Ukrainian friend who would salutorily take her noble leave of me for the evening with Ukrainian for "good night" [na doh-BRAH-nich!]. That *Dobranič!* is recognizable from Russian. Russian actually happens to use a different expression, *pokojnoj noči* "peaceful night," but if Russians did use their words for *good* and *night* as Ukrainians do, it would be *dobraya noč'*–not "doh-BRAH-nich" but "DOH-braya NOACH^{y}." For a Russian, then, mastering Ukrainian is more a matter of adjustment than precisely "learning."

Edward Rutherfurd aptly dramatizes the revival of a sense of Ukrainian as "a language" suitable for writing in his page-turner saga of the history of Russia, *Russka.* In 1827, the poet Karpenko has just recited tales of his Cossack ancestry, and after his friend Ilya suggests that he write them down, he reveals a heretical idea:

> "Actually," he confessed, "what I really want is to write them in the Ukrainian language. They sound even better that way."
>
> It was a perfectly innocent remark: though undoubtedly surprising. "Ukrainian?" Ilya queried. "Are you sure?" Olga, too, found herself puzzled. For the Ukrainian dialect, though close to Russian, had no literature of its own except one comic verse. Even Sergei, always willing to support his friend, couldn't think of anything to say in favor of this odd idea.
>
> And it was now that Alexis spoke. . . . "Forgive me," he said calmly, "but the Ukraine is part of Russia. You should write in Russian, therefore." His tone was not unkind, but it was firm. "Besides," he added with a dismissive shrug, "Ukrainian is only spoken by peasants."

Belorussian is even closer to Russian (in fact its name means "white Russian") than Ukrainian is, and it, too, had status as "a language" as far back as the Middle Ages, when its speakers were administrated by Lithuanians who condoned official business in "White Russian," having no particular stake in elevating Moscow's Russian dialect. Thus geopolitics has elevated Ukrainian and Belorussian as official "languages": but if these regions had continued to be subsumed by Russia and if the Soviet Union had had less interest in suppressing unrest by fostering rather than repressing local speech varieties, then Ukrainian and Belorussian would be the Russian equivalents of German "Mundarts," celebrated mostly by a few local advocates and the occasional books of poetry, folktales, or jokes. Foreigners would tell stories about how "the Russian really gets funky down there in the Ukraine–it was almost a different language!"

Several Languages as Different "Dialects" Finally, just as culture and politics can designate dialects of one "language" as separate "languages," they can also designate languages as distinct as French and Spanish as "dialects" of one. We often hear that a Chinese person speaks the Mandarin "dialect" or the Cantonese "dialect," but in fact the eight main "dialects" of Chinese are so vastly different that they are, under any analysis, separate languages. The Standard German speaker can gradually "wrap his ear around" and "get the hang of" Schwäbisch, but the Cantonese speaker must learn Mandarin as a foreign language. Here is a pair of sentences meaning *I've had my car stolen*:

Mandarin:	Wǒ	bèi rén	tōu le	chēzi.
Cantonese:	Ngóh	béi yàhn	tāu-jó ga	chē.
	I	by person	stolen	car

Taiwanese often speak yet other Chinese "dialects" in addition to Mandarin, which means that most of the Taiwanese immigrants we meet speak two Chinese languages, not just dialects of one.

The reason such different varieties can even begin to be considered "the same language" is because the Chinese writing system uses not letters to represent sounds but symbols to represent whole words. Because the Chinese varieties did all evolve from the same original source, their grammars remain similar enough that they often line up word for word as we have just seen, and this allows the writing system to be suitable for all of the dialects (although because the writing system was developed for Mandarin, there are lacks of fit with the other "dialects").

And then, of course, each of the Chinese "languages" has several dialects, many mutually intelligible only with difficulty. Out in the countryside beyond Beijing, for instance, there are dialects of Mandarin that are as different from the standard as Ukrainian is from Russian and, under other circumstances, could easily be considered "languages" of their own. In Mandarin and other Chinese varieties, single syllables can have different meanings, depending on the tone they are uttered with. In Standard Mandarin, *shu* (pronounced more like "shrew") can mean *uncle* or *book* (among other things), depending on its tone. In the rural dialect of

Wuhan, however, the word for *uncle* is pronounced roughly as "sew" and, as for the word for *book*, shape your mouth to say *Sue* and then say *see.* The tone is no longer the most important distinction between the words, and their shapes are quite different from Standard Mandarin's.

Most Chinese immigrants to the United States in the 1800s and early 1900s spoke a nonstandard dialect of Cantonese, such as the one spoken in the rural region of Seiyap. Educated Standard Cantonese-speaking visitors or immigrants today often have some trouble understanding speakers in America descended from earlier waves of immigration. Thus, rather than having "eight dialects," China actually has several dozen dialects of eight different languages!

Subject Uncooperative: "Language" and "Dialect" and the Nondiscreteness of Language Change

Finally, we might well propose now that, even if cartographic and cultural labels display only fitful correspondence with a conception of "language" and "dialect" based on mutual intelligibility, we might still save these useful taxonomic concepts by supposing that human speech varieties *are* distributed in tidy bundles of mutually intelligible dialects, regardless of how geopolitical and cultural boundaries obscure this. In other words, we might suppose that "in real life," we can just include Macedonian in the "Bulgarian" bundle, ease Moldovan over into the Romanian one, think of Swedish, Norwegian, and Danish as one language, etc., and then everything would be nice and tidy.

But even this doesn't work: in the proper sense, it's not only the geopolitically contingent map that's the problem. Even if the world had not been partitioned into countries, there would still be no intrinsically watertight concept of "language" that would stand up to how human speech varieties are actually distributed on the globe. This follows from the inherently nondiscrete nature of language change, which we already saw producing a "Fratin" phase between Latin and French.

"Where Do You Draw the Line?" Redux: Halfway Between Language and Dialect Because the transformation of a language into a new one is an incremental process, there is a point in this transformation where the new speech variety is clearly akin to its ancestor and other dialects still close to that ancestor, but only fitfully intelligible with them. A speaker of the ancestor or a dialect close to it does not quite process this new one as "a separate language," as a Greek has to learn Hungarian, but then acquiring it takes more than just "making some adjustments," as we would have to do to get along in rural Cornwall of the eighteenth century. In other words, it is common for a speech variety to stand in a relation to another one that is caught between what we intuitively think of as "dialect" and what we intuitively think of as "different language."

The Schwäbisch that threw me in Konstanz is a useful example. It is just one of several "dialects" of German that are so different from the standard that, even for Standard German speakers, becoming able to function in them is almost a matter of learning a new language rather than adjusting to a variation on their own. Swiss German is another one of these "dialects," and shown on the next page are identical panels from an adventure of the French comic character Asterix, in its translations into Standard German, Schwäbisch, and Swiss German.[4] Asterix has been clobbered by a scheming village traitor while standing on guard against invading Romans; the "magic brew" is the village druid's trademark concoction that gives warriors superhuman strength.

Swiss German is particularly instructive. Miraculix's two words *la lige* for "let lie" would be *lassen* and *liegen* in the standard, but *la* is a long way from *lassen*, just as his subsequent *gä* (pronounced "GEH") is from *geben* and his *ke* is from *kein*. In place of the Standard German version's *falsch* for Miraculix's "wrong," the Swiss version has the local *tschärbis* "screwed up," alien to standard. The standard word for "drink," *trinken*, is *suufe*, from a root that in Standard German is used only for "guzzle" or in reference to drinking alcohol in hearty fashion. Germans have to, more or less, "learn" Schwäbisch or Swiss German; regarding another non-Standard German variety, I have heard of Standard German speakers having to take classes in the Kölsch

4. I used to be baffled as to why issues of this formulaic series have sold briskly at newsstands across Europe in dozens of languages for decades (yes, the wordplay but, really, there isn't that much of it in any given episode), but by golly there is something about them that grows on you; I swear that I wish I could spend a month living in that village eating wild boars. *Asterix* does not really work in American English, however–the English translations, done by Englishmen, only begin to work if you imagine them speaking in British accents.

Oh, Miraculix! Do something! Give him some of the elixir that you used against the Romans!	Listen, my good Obelix! I think I left it at the lookout point.	Besides it would be wrong to give Asterix any of it, since everyone who drinks it can't drink any more magic brew without strange things happening to him!

dialect local to Cologne. Although I read Standard German on a regular basis as part of my academic work, when a friend of mine sent me e-mail messages in Swiss German, I found them so opaque that I literally could not grasp even the basic meaning and had to request "translation" into, at least, Standard German if not English.

Still, however, Schwäbisch, Swiss German, and Kölsch are more like other German dialects than like any other languages related to German such as Swedish or English and, after some exposure, one gains a sense of oneself in the "German" orbit in regard to word shapes and grammar. The question, then, is: As a Martian, would you treat Swiss German as a dialect of German or as a separate language if you knew nothing of what speakers call the varieties or where they are spoken? After all, remember that even a Standard German speaker can barely make out anything Schwäbisch, Swiss German, or Kölsch speakers are saying at first.

In making your decision, consider at the same time Spanish and Portuguese. We are accustomed to thinking of them as big, fat, distinct "languages" because they are spoken by formerly geopolitically dominant powers with distinct and rich literary heritages. Yet they are close enough that, if political boundaries had been drawn differently, they would be considered dialects of the "Iberian" language. Spanish and Portuguese speakers can get the gist of each other's spoken languages (although Portuguese have a much easier time with Spanish than the other way around), and I long ago gave up trying to speak, as opposed to read, Portuguese because I find it impossible to keep it separate from Spanish in my mouth and always ended up committing the gaffe of seeming not to realize that Portuguese is not Spanish (which can be a particularly touchy subject in Brazil, where people are weary of Americans assuming that they speak Spanish as other South Americans do). The Spaniard would say *Ese hombre no tiene mis gatos* for *That man doesn't have my cats*, whereas the Portuguese would say *Êsse homem não tem os meus gatos*–the difference here is obviously quite akin to that between Standard and Swiss German. There are plenty of "languages" in, for example, Africa and Asia, as well as ones in Europe such as Italian, whose "dialects" are even more different.

The intersection between language identification and culture gets even messier in cases where group A considers itself to be speaking the same language as group B, whereas group B considers group A to be speaking a different one. On the island of New Britain just off of Papua New Guinea to the east, there are two varieties called Tourai and Aria that to our eyes and ears would appear to be dialects of the same language. That is also the way speakers of surrounding languages feel–they use the same "language" with speakers of Tourai and Aria. But though the Tourai speakers consider what the Aria speak a different language, the Aria consider themselves to be speaking Tourai. The situation gets even more complicated because of the mixing of languages, which has created more than one kind of "Aria." The Aria speakers of the Bolo region, close to where another language called Mouk is spoken, have taken on a great deal of Mouk vocabulary. As a result, the other Aria speakers think of the Bolo Aria as speaking Mouk. However, the Mouk speakers see the Bolo Aria speakers as speaking, well, Aria.

Sometimes cultural distinctions rather than geopolitical boundaries or vocabulary mixture end up fencing off closely related varieties as "separate languages." Senegalese people who speak Mandinka are aware that there are "languages" called Bambara and Dyula spoken in nearby regions but also quite readily mention that they can understand both fairly well. On paper, these three "languages" reveal themselves to be about as close as the various German dialects, and in some linguistic descriptions are treated together as variations on one common theme. The speaker of Anyi in Côte d'Ivoire, if asked about other languages he knows, will usually mention that Anyi and Baule are really "the same thing"–about like Standard English and Scots English, as one Ivoirian told me–but the distinct cultural heritages of Anyis and Baules conditions a sense that the two speak "different languages."

Scots English is, in fact, one of the only ways English speakers experience a variety of their own language that is so different from the standard that it strains the boundaries of what they consider their own language to be.[5] *Auld lang syne*, for example, is Scots for *old long since*. The words are different enough from Standard English equivalents that we usually sing this phrase as an undigested chunk rather than processing the meaning of each word in sequence, and furthermore, even when we know what they mean, there is the further distancing factor that we do not have a set expression "old long since" for "days of yore." This song is the only way most Americans ever encounter Scots; for a healthier dose, here is a passage from the Prodigal Son parable:

> There wis aince a man hed twa sons; and ae day the yung son said til him, "Faither, gie me the faa-share o your haudin at I hae a richt

---

5. Again, creoles are another example: the "English" of many West Indians sits on the dialect/language line similarly; we will look at these varieties in Chapter 4.

> til." Sae the faither haufed his haudin atweesh his twa sons. No lang efterhin the yung son niffert the haill o his portion for siller, and fuir awà furth til a faur-aff kintra, whaur he sperfelt his siller livin the life o a weirdless waister.

Now, we can follow that pretty well, but between far-out versions of words familiar to us like *aince, twa, richt,* and *kintra,* and outright novelties like *atweesh, efterhin, niffert,* and *sperfelt,* this is obviously quite unlike any "English" most of us in America ever hear. Actually, though, political unity with England has gradually brought Scots closer to Standard British English over the centuries. In medieval times, when Scotland was still a separate kingdom, the English dialect of Scotland was well on its way to becoming a separate language, as we see in a snippet from the first fully Scots text, written in 1376:

Thai defendit, and stude tharat,	They defended, and stood there,
Magré thair fais, quhill the nycht	In spite of their foes, until the night
Gert thame on bath halfis leif the ficht.	Caused them on both sides to stop fighting.

This still looks like a sort of "English," more or less, when you look at it long enough, but differences this vast rendered mutual comprehension a dicey affair at best.

Thus what *are* the Schwäbisch and Swiss German in those *Asterix* panels, all considerations of cartography, history, and cultural identity aside–German "dialects" or separate "languages"? If Portuguese speakers can often get the gist of a Spanish news broadcast, in "God's eyes," are Portuguese and Spanish dialects of the same language? Today, there is an influential movement in Scotland to treat Scots as a separate language from English–well, from a linguist's perspective, which side of the line does Scots fall on? Or if it has been inching toward standard English in the past several centuries, which side of the line did it fall on in the Middle Ages, when it was a little farther from the standard than Swedish is from Danish?

The answer, really, is that there is no way to make the call in cases like these. We saw how close dialects can be compared to "covers" of an original song. A case like Swiss German brings to mind an episode of *The Simpsons* lampooning *Mary Poppins,* complete with song parodies of "A Spoonful of Sugar," "Feed the Birds," and others. The songs did not use the melodies from the Disney movie, Weird Al Yankovic–style, but were specially crafted with basic shape, rhythm, and harmonic flavor paralleling the originals just enough to instantly recall the songs parodied without sparking a lawsuit.[6] Swiss German stands in a relation to Standard German analogous to that between these song parodies and their models.

If Belorussia and the Ukraine were still regions of Russia instead of separate countries, then Belorussian and Ukrainian would present the same conundrum as Schwäbisch and Swiss German. Ukrainian is definitely not Russian–but then it's more like Russian than like any other language, and enough like it that I could alternately entertain and annoy my college friend by "making up" Ukrainian words based on Russian. Cases like this show that speech varieties differ from one another along a continuum, on which no definite signpost can be placed distinguishing where "dialect" stops and "language" begins.

One "Language" Bleeds into Another "Language" The final reason there is no such thing as a "language" in any intrinsically coherent sense is because, in many cases, one runs into another one just as green is neither yellow nor blue, but a mixture of the two. An example is a group of dialects called Gurage [goo-RAH-gay] spoken in Ethiopia, one of several languages related to Arabic and Hebrew spoken in that country. Does anyone still play that game "Telephone" where people sit in a circle and one person whispers something in the next person's ear, and then that person, having heard something slightly different from what the first person said, whispers what she heard in the next person's ear, and so on, until what comes out on the other end is hopelessly different from the original sentence? A similar phenomenon occurs with what is called a *dialect continuum.* The way to say *He thatched a roof* in Gurage dialects differs slightly from one region to another (the ə is called schwa, the sound of *a* in *about*):

---

6. Hats off to *The Simpsons'* house composer Hans Zimmer, who also composed what I consider the best theme song in the history of television, for the late, great *The Critic,* luckily still shown on Comedy Central.

Soddo:	kəddənəm	Chaha:	khədərəm
Gogot:	kəddənəm	Gyeto:	khətərə
Muher:	khəddənəm	Endegen:	həttərə
Ezha:	khəddərəm		

Soddo and Gogot have the same word; Muher only differs in having an initial *kh* instead of *k*. Ezha makes the small change from here of substituting *r* for *n*; although this is cake for Muher speakers, it's already pretty odd for Soddo speakers. Chaha and Ezha are quite close, but then Gyeto changes the *d* to a *t* and lets go of the final *m*. Neither of these sound-change processes is at all unusual, but to someone who grew up on *kəddənəm*, *khətərə* is almost incomprehensible at first. Endegen substitutes *h* for *kh*, a natural little jump, and doubles the *t*. Endegen's *həttərə* is so different from Soddo's *kəddənəm* that Endegen (and Gogot and Muher) speakers process Soddo as essentially a different language. Thus what is "Gurage"? Relationships among many of the varieties are what we think of as "dialectal," but just as many relationships are akin to that between Spanish and Italian. Gurage is neither a bundle of "dialects" nor a bundle of "languages"–it is a conglomeration of varieties related to one another to various degrees.

This is not rare; linguists encounter dialect continua all over the world, often linking what are conventionally known as separate "languages." In the Central African Republic and the former Zaire, from region to region there are various "languages" that differ from one another only to the extent that we would imagine of "dialects." Here is how to say *Me, I'm going to the village to build a house* in seventeen of these languages:

Bobangi:	Ngai, nakoke o mboka notonga ndako.
Nunu:	Ngai, namoke o mboka notonga ndako.
Libinza:	Ngai, nakakende o mboka nakatonga ndako.
Lusakani:	Ngai, namoke o mboka notonga ndako.
Mpama:	Ngai, nakei mboka nakatonga ndako.
Liboko:	Ngai, nakei o mboka nakatonga ndako.
Loyi:	Ngai, nakei mboka natonga ndako.
Impfondo:	Ngai, nakei o mboka mpfoa ya itonga ndako.

Enyele:	Nga, nakei mboka botonga ndako.
Bomitaba:	Nga, nakei mboka eke otonga ndako.
Likuba:	Ngai, nasoke mboka otonga ndako.
Likuala:	Nga, nake o mbowa notonga ndako.
Moyi:	Ngai, nakeke o mboka notonga ndako.
Mboshi:	Nga, izwa mboa otonga ndai (ndao)
Koyu:	Nga, lizwa mbooka etonga ndako.
Makua:	Nga, ikendi mboga etonga ndago.
Bongili:	Ngai, nake mboka na kotonga ndako.

One could make a similar list of identical sentences in the various German "dialects," and they would often differ more than these "languages" differ from one another.

Turkish is one of several Turkic languages, many of which are highly similar to one another, such that, in many parts of the Turkic-speaking region (including many of new "stans" freed from the former Soviet Union), one "language" bleeds into another one through intermediate dialects. Here, for example, is *eight* in seven of these languages going from west to east:

Turkish:	sekiz	Kazakh:	segiz
Azerbaijani:	səkkiz	Kirghiz:	segiz
Turkmen:	sekiz	Uighur:	säkkiz (here, ä = the *a* in *cat*)
Uzbek:	sakkiz		

You couldn't pay the Romance languages to match up that nicely. Even though their relationship is clear on all levels, here's what happens to the word *eight* in seven of them:

French:	huit	Romanian:	opt
Spanish:	ocho	Occitan:	uèch[7]
Italian:	otto	Catalan:	vuit
Portuguese:	oito		

7. I like that one, too.

There is not a continuum of mutually intelligible dialects across all seven of the Turkic languages; there is one between Turkish and Azerbaijani, as well as between some other pairs. In general, however, all of these "languages" are closer than are all of the "dialects" in many "languages." Turkey, Turkmenistan, Kirghyzstan, Uzbekistan, Kazakhstan, and Azerbaijan are often referred to, as I did earlier, as united by being "Turkic speaking," with an implication that they all speak in some sense one "Turkic" language, even if that language is not Turkish proper. The roots of this concept of a "Turkic" hovering somewhere between "language" and "dialect" lie in the fact that a general "Turkic" system varies incrementally from one region to the next, confounding any attempt to apply taxonomic labels in any consistent way.

Predictably, Serbs and Croats have been known to treat "Serbian" and "Croatian" as different languages and even often claim to have difficulty understanding one another. As with the Romanian-Moldovan case, writing lends an artificial sense of distinction: Serbian is written in Cyrillic, whereas Croatian is written in the Roman alphabet. There are also, as always, some differences in vocabulary. Yet traveling from humble hamlet to humble hamlet across the former Yugoslavia, apart from the artificial division created by writing and cultural conflict, the linguist encounters a continuum of dialects changing Gurage-style from village to village. Among immigrants from the former Yugoslavia today it is quite common to see couples, one member Serbian and the other Croatian, conversing easily in a single language, Serbo-Croatian. Culture and politics make the call between dialect and language here and have continued to do so–after the Dayton Accords, a dictionary of the "Bosnian" language was published.

In larger view, this particular continuum encompasses "languages" even beyond Serbo-Croatian. Not only can Bulgarians understand Macedonians next door, but Macedonians on the border with Yugoslavia can communicate with Serbo-Croatian speakers on the other side, such that Serbo-Croatian, Macedonian, and Bulgarian form a grand continuum. Standard Serbo-Croatian and Bulgarian are as different as Spanish and Italian but are linked by a procession of dialects–and even a whole "language"–falling on a continuum linking them in a kind of living exhibit of one morphing into another in space just as languages morph into one another in time.

There are cases in this region where not only the sound of a word but its very meaning changes incrementally as well.

Dalmatian coast:	vridan	"industrious"
Bosnia, Montenegro:	vrijedan	"industrious"
Serbia:	vredan	"industrious"
Macedonia, Western Bulgaria:	vredan	"industrious" or "harmful"
Southeastern Bulgaria:	vraedan	"harmful"

The question arises, then: If all of these dialects were spoken in some uncharted region rather than artificially corraled into "countries," where would you draw the line between one "language" and another one?

Intelligibility: Taxonomic Quicksand One manifestation of this "neither fish nor fowl" aspect of many varieties is that dialects that are in essence extremely close can still be just barely mutually intelligible. This results from small but sharp differences in the sound systems and from differences in the semantic evolution of vocabulary. For example, though Danes, Swedes, and Norwegians can converse, intelligibility is not all peaches and cream. Just as among Romance languages French has transformed the original material more than most of the others (from Latin *cantare* "to sing," Italian still has *cantare*, Spanish has *cantar*, but French has *chanter* [shaw^{ng}-TAY]), Danish is the "advanced" one in Scandinavian. For "to play," Swedish has *leka*, and Norwegian has *leke*, both pronounced approximately "LEH-kuh," but Danish has *lege*, whose archaic spelling masks that it is pronounced "LIE-uh." This means that, for Norwegians and Swedes, getting just what words a Dane is saying is a bit of a strain.

On the other hand, between Norwegians and Swedes, the similar sound systems make understanding what words are being used unproblematic, but problems arise because of different *meanings* of the same root: Norwegian *rolig* is "calm," but in Swedish the same root has drifted to mean "funny," just as *silly* drifted in English from "blessed" to "idiotic." *Dyrke* is "to cultivate" in Norwegian and "to worship" in Swedish; Norwegian's *bløt* is "soft," Swedish's *blöt* is "wet"; Norwegian's *tilbud* is "offer," Swedish's *tillbud* is "accident." Because Norwegian was still "Danish" more recently than Swedish was, Danish tends to have the same meanings as Norwegian, and thus Swedes and Danes have the same problem. Because Danish is the odd one out in regard to sound system, whereas Swedish has gone its own a way a bit in regard to word meanings, it has been said that "Norwegian is Danish spoken in Swedish"–that is, Norwegian, which parallels Danish's word meanings, is how Danish would come out if its sound system weren't so independently minded and were therefore more like the Swedish one.

This kind of ambiguous degree of intelligibility exists worldwide between speech varieties that look highly close on the page. Oruqen and Evenki are two closely related–well, linguists don't know what to even begin to call them–that straddle a border between Russia and northwestern China. Line the two up on the page and they look as close as the Turkic varieties listed earlier. But in real life on the ground, the intelligibility matter is tricky. In line with what we would expect from what they look like in print, Oruqens often claim to be able to speak with Evenkis–but then have been shown to not be able to understand a tape played of Evenki being spoken. Much of the problem appears to be that, in Oruqen, accent always falls on the last syllable of the word, whereas in Evenki it can fall in various places as in English. Evenki has *óllo* for fish, Oruqen has *oló*; and so on. When this difference is applied to every word in the language, the cumulative effect is considerable–imagine if we had to communicate with someone who, in addition to having what we processed as a thick accent in general, said things like "Venn this diffRENS iss ibLIED to avREE wirt in dah lanGWIDGE, dah camalaTIFF EFfect iss cansaderaBULL." This would still be "English"–it sure isn't German; but you'd almost wish it *were* German so that you could claim not to speak the person's language!

Thus even the intelligibility issue is messy: any metric of intelligibility one tried to fashion would trip up on the fact that intelligibility and taxonomic closeness do not walk in anything approaching a lockstep. Certainly dialects that are not close on the page will also *not* be mutually intelligible–but then, when they *are* close on the page, they may or may not be mutually intelligible.

Linguists are often asked, "What's the difference between language and dialect?" Often, the answer they give is that it is a "difficult question," which depends as much on culture, history, and politics as on linguistic reality. This response, however, refers only to *language* and *dialect* as labels. The linguistic reality does not lend itself gracefully to *any* underlying conception of a language/dialect distinction. The geopolitical and cultural factors only make clearer a problem that would exist even if there were no such subdivisions and humans simply coated the earth in little hunter-gatherer bands as they once did in Paleolithic times. Properly, the language/dialect distinction is, in the pure logical sense, meaningless.

Certainly there are "languages" with only one dialect, such as many languages spoken by only a few hundred hunter-gatherers in places like Papua New Guinea. And certainly there are bundles of closely related dialects where none stray particularly far from the basic template. Korean has its dialects (predictably, for example, North Koreans speak markedly differently from South Koreans), but all are readily mutually intelligible. Because Modern Hebrew is spoken in a tiny country, and only has been so for less than a century, there are no spoken dialects of it that strain an Israeli's sense of what "Hebrew" is. But the crucial point is that this is by no means the "default" situation: on the contrary, if anything, these situations are the exceptions, more typical of smaller groups of speakers. Typically, what looks from the air like "a language" is actually a much hazier business on the ground. Korean is relatively uniform, but Japanese speakers in Tokyo can barely follow speakers from the Ryukyu Islands, and on the page the Ryukyu dialect is about as far from Standard Japanese as Schwäbisch is from Standard German. Hebrew is pretty tidy, but its neighbor and relative Arabic differs so much from country to country that the various "dialects" differ about as much as Spanish, Italian, and Portuguese do from one another. Thus though there are cases where speech varieties *happen* to fit into an idealized language/dialect template, they usually do not. As such, properly speaking, there are no "languages."

If it is possible to save any remnant of our terminology, the best we can say is that there are innumerable *dialects* in the world, related to each other to various degrees, sometimes clumping into complexes particularly close to one another, but generally not so close that all are mutually intelligible, with distances often so great between some of them that their speakers do not consider themselves to be speaking "the same thing" in any sense.

It's all about dialects, then: language change has split the first language into tens of thousands of dialects that we arbitrarily group into "languages" according to approximate notions of intelligibility and the dictates of the cultural and political developments of the moment. Dialects are everywhere–and always have been: even Old English had them. I oversimplified a bit in depicting the

Angles, Saxons, and Jutes as speaking one original "language" when they invaded England. Because these peoples had lived in separate places in each of which the unwritten West Germanic ancestor of English had developed in slightly different directions, they spoke at least three dialects of Old English, as we see from three renditions of the first line of the Lord's Prayer:

West Saxon:	Fæder ūre, thū eart on heofonum
Northumbrian:	Fæder ūrer, thū art on heofonu
Mercian:	Feder ūre, thū eart on heofenum

Thus the dazzling variety among British dialects stemmed from at least three slightly variant founding *dialects,* not a single variety.

We would further predict that there would even be dialects of dialects, and this is exactly what happens, which in turn highlights once more that the original language has developed not into just six thousand more "languages" but, more properly, into tens of thousands of variations on variations corresponding to the tens of thousands of speech communities on the earth, obviously vastly outnumbering the mere six thousand "languages" that we can approximately delineate. When I mentioned the Bavarian German translations of *Asterix* to a Bavarian in Germany, her first response was a good-natured complaint that they had not translated into *her* dialect of Bavarian, and she proceeded to give me some of the differences between her speech and that depicted in the books. A few weeks later in America I mentioned the Swiss German translations to a Swiss person, and he immediately said, "Well, the one problem is that of course they didn't translate it into *my* dialect!" The word for "messed up" in the Swiss German *Asterix* excerpt, *tschärbis,* is from only one dialect of Swiss German, not used by all German-speaking Swiss. Dialects is all there is.

Two Tongues in One Mouth

There is a nuance to be added to our developing picture of human speech across the globe. Just as many people in the world are bilingual in two or more "languages," a great many people control more than one dialect of a language. In particular, it is common for members of a community or society to speak both a standard dialect and a nonstandard one, especially today with the spread of education and the centralization of economies lending increasing numbers of people more contact with the standard dialect than was the case in earlier periods of history.

The classic example is Swiss German. For German-speaking Swiss, Swiss German is the language of the home, the language learned first, the language of the casual, the familiar, the intimate–and for *all* Swiss rich and poor; Swiss German is not a class issue the way, say, Appalachian or Black English partly are. Standard German, however, is the language of writing, official announcements, and all scholastic endeavor. All students are taught in Standard German, its grammar is the one taught in the schools, and it is used almost exclusively of Swiss German on the radio and on television. Thus all German-speaking Swiss by the time they are mature speak and understand both Standard German and Swiss German, thinking of them as variants on a common theme, each with their particular sphere of appropriateness. My expounding about Swiss German, dutifully giving samples from it on paper, contrasts with its actual "place" in its speakers' consciousness, which is as an integral but strictly informal part of life, usually only seen in writing in personal messages in local newspapers or in personal letters. At the Frankfurt airport I saw a man, apparently Swiss, chuckling and hooting while reading an *Asterix* edition in Swiss German, it being funny to him to see a book written in the dialect.

There are similar situations throughout the world, and they are called *diglossia,* from the Greek for "two tongues." The Arabic an Egyptian of any class speaks at home is actually a different language from the Modern Standard Arabic used in writing and in scholarly instruction. Just as the Swiss German speaker has *suufe* at home and *trinken* in print, the Egyptian refers to his nose as a *manaxir* but would write *'anfun* in print. This is also true of different non-Standard Arabics spoken in Morocco, Algeria, Nigeria, Sudan, Syria, Lebanon, Saudi Arabia, etc. I once heard an educated Moroccan journalist describe his childhood saying casually that he had spoken "Moroccan" at home and then learned "some Arabic" in school, neatly demonstrating that, in the mind of a Moroccan Arabic speaker, Modern Standard Arabic is not just a hoity-toity way of speaking what he learned at his mother's knee, but essentially a different language that must be carefully taught. I also saw three Finns unable to agree on just how to say "Hey, look–a shortcut!" in large part because of dialect differences; one of them came from a region where the local dialect is different enough to inspire affectionate, locally produced jokebooks just as Schwäbisch and other dialects do in Germany. Yet all of them were fully functional in Standard Finnish as well. The standard dialect of Indonesian used in writing and taught in books is a scholarly creation designed to parallel Western languages in its grammar as much as possible; in their everyday lives, Indonesian speakers use an array of nonstandard varieties that are often only fitfully intelligible with the standard one. The layers of language in Javanese that we saw on page 50 are another example of diglossia (although in that case there are actually "middle class" forms as well, such that we are really dealing with triglossia, something else not unheard of worldwide).

Diglossia is a manifestation of a hierarchy of social domain that all human speech varieties observe to some extent; diglossia goes to the extreme of dividing the labor between two distinct grammars, but in many languages the same kind of distinction is indexed through vocabulary alternates and various set expressions. There is a song that longtime residents of Oakland, California, sometimes sing dedicated to the city, and one of the lines goes, "Where did all the people go when Frisco burned?/They all came here to Oakland and they never returned!" That line has always struck me, overthinking such things as I tend to, as a bit "off" in tone. It's that word *returned*. The song has a red-blooded, rah-rah feel that leads you to expect it to end with the spelling out of O-A-K-L-A-N-D and a rousing "Whoo!" or the like (although it doesn't, actually) and, as such, *returned* is too formal. We say *went back* or *came back* in casual English; *returned* is for writing and formal situations. Imagine asking your significant other, as you peel the potatoes, "When are you returning tonight?"–you'd either be (1) new to English, (2) striking an irritated or ironic tone, or (3) very, very strange. Our diglossia splits between *come back* and *return*, or *check out* and *examine*, or *kids* and *children*.

We can see an illustration of how diglossia plays out when distributed across two dialects with one more look at our friend Asterix, this time from the Bavarian German edition (see page 90). Asterix and Obelix, mistaken as Goths the Romans are chasing (don't ask), are disguised as Romans to throw them off the scent, and Asterix instructs Obelix on how to greet any Romans they

. . . and listen, Obelix: if we meet Romans, you are Obelus the private and I am Asterus. And say, "By Jupiter" and "Ave."

Hee, Hee! This is fun!

Pull yourself together! Soldiers are coming!

Ave, Comrades! Have you seen anything of the two Goths?

Ave and By Jupiter . . .

Excuse us, but my friend Obelus is very happy today.

Lucky him! Even looking for frightful Goths he still has fun.

meet. The translator smartly has the Romans speaking the "official" dialect Standard German, while Asterix and Obelix speak Bavarian among themselves; but then, like almost all Bavarians, Asterix is diglossic between Bavarian and the standard: when speaking to the Romans, he switches to the standard:[8]

There was a period when educated subjects of many parts of the Roman Empire were diglossic in what were, at the time, local dialects of Latin and Latin itself–the pedant's complaint on page 40, for example, shows that people raised on "Fratin" learned Latin through tutelage in school (if they were among the few who went to school), just as Swiss German speakers today learn Standard German in school.

Recall, though, that as late as the 1700s speakers of nonstandard "French" dialects did not have any appreciable familiarity with the standard one; Caxton's *eggs/eyren* story shows the same thing in medieval England. Diglossia can be acquired through religious instruction–the centrality of the Koran in Islamic life has long ensured at least a basic familiarity with Standard Arabic even among uneducated Muslims, for instance. However, education spreads diglossia even more widely and drills it harder. Until a few centuries ago, education was a relatively elite privilege in all societies; today, however, as education reaches ever more people worldwide (although surely not nearly enough), diglossia is approaching the status of a norm and will certainly increase in this century as new generations in previously isolated regions increasingly cast their lot with the outside world.

If you open up one of those bags of little plastic dinosaurs, you usually get about six "kinds": a *Tyrannosaurus,* of course; a *Stegosaurus* (the one with the plates on its back); *Triceratops* (the

---

8. Just how communication is supposed to be playing out between the Gauls and the constantly encroaching Romans is not quite clear in the *Asterix* series. The Gauls are supposed to be speaking Gaulish among themselves, a Celtic language related to Irish and quite unlike the Latin the Romans are supposed to be speaking. And yet the Gauls and the Romans seem to have no trouble communicating, despite it being specified (elsewhere in this episode, for instance) that the Romans know no Gaulish and it never being indicated that the Gauls are switching to some form of Latin when speaking with the Romans. For the record, in this episode, the Goths speak Standard German, but always written in the Fraktur "Gothic" alphabet, which we associate with newspaper titles.

horns); a *Brontosaurus;*[9] and so on. But if you really get into dinosaurs you see that each one of these standard *Flintstones* dinosaur types was actually one of an almost numbing array of variations on a theme. There were lots of kinds of stegosaurs–this one had spikes down half its back instead of plates, that one had smaller plates overall, etc.; there were lots of horned dinosaurs, one kind with one horn on its nose, another kind with a bump on its nose, still others with horns all around the frill; *Tyrannosaurus* was one of a couple of dozen similar but slightly variant creatures discovered worldwide; there were runty-sized brontosaur types, and some of them had armor in their skin; etc.

This is what "languages" are like. The dinosaur parallel goes even farther in that similar dinosaur types often lived together in the same areas: the carnosaur *Allosaurus* shared its environment with the smaller *Ceratosaurus,* which had a little horn on its nose; the duckbill *Corythosaurus* with the helmet decoration on its head and *Parasaurolophus* (the one with the marvelous curved tube flying backward off the top of its head) lived in the same places as well. Evolution produced not only "types" but subtypes of dinosaur (and even these subtypes had differing species within them); language change has developed "subtypes" of language, not just occasionally–as we are sometimes misled into thinking by statements such as "Italian has a lot of dialects"–but almost always. Dialects are the norm; dialects are what happened under typical conditions. It's all about dialects. For our purposes, forget "languages"!

So far, we have seen how the first language developed into turtles and cats, and then we looked a little closer and saw how it developed into snapper turtles and sea turtles; Burmese cats and Siamese cats. Now we will take the next step and see how the first language also developed quite often into mules and, well, my cat.

---

9. Dino fans: Yes, I know, but really, how many other people know what an *Apatosaurus* is?

3

The Thousands of Dialects Mix with One Another

So far, in describing how the first language split into thousands of subvarieties, I have implied that speech varieties have developed like a bush, starting from a single sprout and branching in all directions, each branch then developing subbranches, and so on, culminating in a dense web of a plant whose outer layer is crowned with thousands of leaves, symbolizing languages (or dialects). Allowing that the bush analogy cannot capture the fluid nature of the degree of relatedness between dialects, we could think of leaves lining the same twig as closely related varieties such as the German dialects, the leaves over on the next branch as the English dialects, and branches way over on the other side of the bush as, perhaps, the languages of Polynesia.

Indeed, linguists who study language change and the family relationships between languages have traditionally taken the "family tree" model as central in how language has developed. Yet particularly in the past twenty years, language change and classification specialists have come to realize that this model actually only takes us so far in describing the reality of what has really happened to the first language as it has spread across the planet.

Just as it is inherent to languages to change gradually into new ones, it is equally inherent for them to mix with one another. Moreover, just as language change is an unbroken process along which no lines of demarcation can be drawn, language mixture is along a continuum of degree. Viewed close up, not only are "languages" clusters of dialects but, in the past 150,000 years, the dialects of languages have been constantly adopting words, sounds, and sentence structures from neighboring dialects of other languages, spoken by peoples encountered through migration, peoples who migrated into the region themselves, or, especially lately, peoples imposing their dialect of a language through communications technology and educational channels.

The result is that we can only preserve our family tree or bush analogy by making it a flowering bush, with each dialect not a leaf but a flower, with the various flowers cross-pollinating one another. However, we might do just as well with another analogy, say, stew–a good spring lamb stew without much juice (because the juice messes up the analogy). Clearly, one can distinguish the lamb from the peas from the potatoes from the carrots from the leeks from the rosemary leaves. Yet all of these ingredients, if it's a good stew, are suffused with juice and flavor from the other items as well. Every now and then, you even encounter a piece of something that, covered with liquid and cooked out of its original shape and consistency, you have to work to figure out the original identity of, especially if you are one for tossing all sorts of things at hand into the stew–green beans, chickpeas, turnip cubes, shoelaces. Overall, there is nothing in this stew that tastes or even looks like it would if you had just dumped it into a pot of boiling water by itself. That's the very nature of stew, and that's the very nature of the thousands of offshoots of the first language today.

Language Mixture Level One: Words

To see a prime example of how central mixture is to the life cycle of a human language, this time we do not need to go as far afield as Mandinka or Evenki, and not even to French or German. As it happens, English itself has a stunningly bastard vocabulary.

English: An Illegitimate Vocabulary

When we check out a word's history in a dictionary, we are accustomed to finding that the word traces to Dutch, Spanish, Greek, French, and other languages. For the word to trace to Old English is but one of many possibilities, and not even the usual one. What we rarely have reason to consider is that this is not the experience of speakers of all languages when they consult a dictionary. There are no languages or dialects in the world that have not undergone influence of some kind from others, but in a great many cases, on a trip to the dictionary, finding out that the word began as an import is not the virtual default as it is for us. Poles, for instance, find that most of the basic words in their language are traceable not to Dutch, Spanish, Greek, and French but to the Proto-Slavic ancestor it descended from. The equivalent is true of Chinese, Arabic, and many other languages with a written history.

Out of all of the words in the *Oxford English Dictionary,* however, no less than ninety-nine percent were taken from other languages. The relative few that trace back to Old English itself are also sixty-two percent of the words most used. Therefore authentically English roots, such as *and, but, father, love, fight, to, will, should, not,* and *from,* are central to speaking English. Yet the vast majority of our vocabulary originated in foreign languages, including not merely the obvious "Latinate" items like *adjacent* and *expedite*, but common, mundane forms not processed by us as "continental" in the slightest.

For example, every single word in that last sentence longer than three letters originated outside of English itself, despite the fact that the sentence, though hardly street-corner casual, is not exactly bristling legalese either. In fact, it's pretty easy to "cook up" that kind of sentence. What would be harder is to come up with one made up only of words that come from Old English. In fact, that last sentence was one, but I only got away with it by relegating the word *sentence* to the sentence before, and I also found it impossible to express the *full* point I intended without a French or Latin word. Namely, I would add that it would be hard to come up with all-English sentences *of any use*, even though one could play the game of composing a shortish, choppy, rather bland text by using only the Anglo-Saxon core words. Only by making up new English words in combining the roots in new ways would we be able to get beyond this, as some writers have tried through the centuries (*speechcraft* for Latin's *grammar*, for instance). On the other hand, if one were given the dispensation of using the small collection of Anglo-Saxon grammatical items like *and, but, will, if, should,* and *would* but otherwise forbidden to use any other English-derived words, then one could compose entire books in language that would seem only slightly ornate.

Thus English's vocabulary is like San Francisco's architecture: thriving and beautiful but with ultimately sparse roots, with only a few of the buildings constructed before the earthquake and fire of 1906 still standing. English lost most of its original vocabulary through three main lexical "earthquakes."

Vikings invaded and settled in the northern half of Britain starting in 787; they spoke Old Norse (ancestor of today's Scandinavian languages) and scattered about a thousand words into English. They were not merely "cultural" terms but staples like *both, same, again, get, give, are, skirt, sky,* and *skin.* If I tell you that on a foggy Thursday, a sly, dirty-necked, scowling outlaw skulked into the bank with a knife, ransacked it, and crawled out the window seeming happy, every word came from these Vikings except *a, into, the, with, it,* and *out.*

Then, in 1066, French speakers took over England for roughly the next two hundred years. Actually, these "French" people were Vikings again, having taken over northwestern France and switched to French over the generations; their ancestry was why these French were called the Normans–that is, Norsemen. They introduced no fewer than about seventy-five hundred words, as well as about twenty-five hundred that no longer survive. The sheer volume of thoroughly ordinary "English" words that originated in France is stunning: how "French" do words such as *air, coast, debt, face, flower, joy, people, river, sign, blue, clear, easy, large, mean, nice, poor, carry, change, cry, move, push, save, trip, wait, chair, lamp, pain, stomach, fool, music, park, beef, stew, toast, spy, faith, bar, jail, tax,* or *fry* feel to us today?

The "Latinate" layer most perceptible to us as a word class apart came after the withdrawal of the French, with the increasing use of English as a language of learning–hence *client, legal, scene, intellect, recipe, pulpit, exclude, necessary, tolerance, interest,* et alia (including *et alia,* of course).

All of this means that an English that had developed without these lexical invasions would be incomprehensible and peculiar to us. The Beautiful People would be *Scīene Lēode* rather than the French words we use today;[1] *conscience* would be *inwit* "knowledge within"; a succession would be an *æftergengness,* an "aftergoing." There is something comforting in the idea of our having words like

1. German speakers, note the similarity here to *Schöne Leute,* to give an idea of how similar to German English would be if history had proceeded otherwise.

inwit and *æftergengness*–as English speakers, we have gotten used to a great many of our compound words being essentially opaque, such that we have to learn them by rote, but wouldn't it be nice if most of our "big words" made at least some immediate sense to us because they were composed of roots drawn from the ordinary level of the language? The word *conscience* to us is basically a unitary piece of stuff–we know that the *-ence* is an ending usually making an adjective into a noun, as in *magnificent / magnificence*, but the *consci-* part is something we just have to "know." But to the Ancient Roman, *conscientia* was perceptibly composed of *con-* "within" and *scientia* "knowledge."

The Practical Pluses and Minuses of Bastard Origins

Speakers of languages closely related to English that did not undergo such lexical replacement from French and Latin still have this luxury. A friend of mine envies the Germans for having a language that remains "transparent" to them, and what he means is that German for *conscience* is *Gewissen, wissen* meaning "to know" and *ge-* nicely making it a noun; *succession* is *Reihenfolge* "row-following"–what could be easier? To be sure, semantic drift over time ensures that affixed German words are often not this easy to parse. Yet even these meanings are usually relatable to their roots with a bit of imagination, whereas we have no such luxury with words like *commit*. First of all, even if we look at other words with the *mit* root, its meaning appears highly fuzzy at best: *admit, demit, transmit*; furthermore, what does the *com-* mean? English just swallowed these "mit" words whole from Latin, where, for the record, *mittere* meant "to send," which helps somewhat with *admit, demit,* and *transmit*. But how were we to know that?

On the other hand, our French and Latin inheritance does give us a certain advantage, although even this in practice ends up being a mixed blessing. English is so larded with Latin and French words that we have a good head start on learning the vocabularies of French and other languages descended from Latin. This is especially true of the more formal layers of these languages. Most of English's French and Latin words percolated in "from above": Latin words were incorporated in the realm of the learned (*adjudicate, anatomy*), and the Norman French, as rulers, were especially influential in prestigious realms such as government, the law, and the arts. Thus, today we have doublets like *pork* for the New White Meat, but *pig* for the humble, snorting creature the meat comes from, whereas the Germans have no trouble calling pork *Schweinefleisch*. *Aid* is a more lah-di-dah word for the original English *help*, *commencing* is a fancy way of *beginning*, an *aroma* is a buttoned-up version of a *stench*, and so on. This means that the hardest part of acquiring a vocabulary in French is at the beginning, mastering the basic words for which we retain the English originals–*pain* for *bread*, *eau* for *water*, *poisson* for *fish*. After we've got the heart of these, though, we can follow a French newspaper article pretty well without having been explicitly taught the more formal words like *association, opportunité,* and *présent*. This is not because these were the original English words but because we inherited them from Latin and French hundreds of years ago. Indeed, we have so many of these Latin/Romance words that in Russia, where there is currently a revival in the teaching of classical languages, one benefit classics teachers point to in learning Latin, echoed by students, is that learning it gives a Russian a leg up on learning English!

Yet this Latin/French legacy can also leave English speakers somewhat frustrated when it comes to learning languages other than the Romance ones. Americans who found that after a few years of learning Spanish they could read a Spanish newspaper fairly comfortably are disappointed to find that after a few years of German classes, they can get at best the gist of even a rather simple magazine article or comic book, whereas an issue of a prestigious newspaper like *Die Welt* is virtually unintelligible. A person can even become fluent in casual spoken German after a year in the country and still be barely able to get through a page of Germany's general interest magazines such as *Der Spiegel* or *Stern*. This is because the massive parallels between the "high" vocabulary in English and the Romance languages are due to a historical accident: most languages' vocabularies do not match to this extent on any level.

German is especially deceptive here because, given that it is closely related to English, many of the *basic* words are quite similar to ours, unlike in French: instead of *pain, eau,* and *poisson, Brot, Wasser,* and *Fisch*. But once we get to associations, the present, and opportunities casually discussed in *Der Spiegel,* the words are *Verband, Gegenwart,* and *Gelegenheit*–this means that one has to devote as much attention to internalizing the high layer of the German vocabulary as one does to the everyday layer in French. And then, when we venture outside of Germanic or Romance, both layers are equally opaque for us: Russian's bread, water, and fish are *xleb, voda,* and *ryba;* association, opportunity, and the present are *soedinenje, vozmožnost',* and *nastojaščee.*

Our bastard vocabulary also deprives us of an experience quite ordinary to speakers of most other languages in the world. We have seen how Spanish and Portuguese are close enough that if spoken in the same country, they might well be considered dialects of one language. Zulu and Xhosa in South Africa are similar. Dutch and German are just a notch or so farther apart than the previous two pairs; Serbs and Bulgarians can manage basic communication with effort. Even when any kind of conversation would be impossible, for most people in the world, there are languages whose mastery would be essentially a matter of getting used to a variation on their own. A Hindi speaker would have pretty smooth sailing learning Gujarati, Marathi, Bengali, and various other related languages; the Southeast Asian villager who travels to neighboring areas will often find a language that feels like a fun-house-mirror version of his own.

We English speakers, however, do not have this advantage with any language on earth. Our vocabulary is so shot through with foreign loanwords on all levels that there is no language whose basic vocabulary is mostly akin to ours. If the Norse and Norman invasions, as well as the more cerebral but equally decisive Latin one, had not occurred, then it would be Dutch and the Scandinavian languages that we could pick up as easily as Russians do Polish, and German would be just a bit more of a stretch.

But history had it otherwise. The languages spoken today where the Angles, Saxons, and Jutes migrated from, Dutch and its close relative Frisian, remain the languages closest to English today. Indeed we can make English and Frisian line up by the contrivance of a sentence like "good butter and good cheese is good English and good Fries," and I do not need to translate Dutch fragments like *Mijn naam is John* or *Wat is dat?* But the likenesses do not go far beyond the likes of this, and an English speaker has almost nowhere to grab onto in a passage of ordinary Dutch like *In elk McDonald's restaurant is een speciale medewerker aanwezig om u te advisieren, vragen te beantwoorden of gewoon te helpen*—"In each

McDonald's restaurant, there is a special assistant present to advise you, to answer questions, or just to help."[2]

Thus our lexical "earthquakes" left us with a boost in acquiring one level of the vocabulary of the Romance languages, but that's it. A language's vocabulary is a vast and multilayered affair, highly underrated in relation to its centrality to really mastering a foreign tongue. For most humans, there is at least a language or two that gives them an exemption from having to tackle this head-on, but English's peculiar history deprived us of having a head start on *all* of the vocabulary of *any* other language.

The Myth of English's Inherent "Flexibility": Language Mixture Anticipated Globalization by Millennia

Thus the story of English includes not only the changes in its sounds and grammar that transformed Old English into the language I am writing in, but at the same time rampant vocabulary mixture with other languages. Writers often attribute this to English being particularly "flexible," but this is a post hoc misconception. Any language can incorporate boatloads of foreign words, and most have. All it takes is contact between cultures, and of course there is no culture on earth that does not have, or has not in the past had, significant contact with other peoples. Indeed, intercultural contact is the very heart of human history, and thus the six thousand human languages are replete with the results of it.

Generally speaking, when a dominant proportion, or often just a highly influential minority, of a language's speakers become bilingual between that language and another one over a long period of time, the gates open for exchange of words of this kind, usually flowing more from conqueror to conqueree than in the other direction. Cultural disposition makes some languages more resistant to borrowing words than others, to be sure, but the space to maneuver here is pretty narrow. No language is completely immune to borrowing, and when cultural contact becomes particularly heavy, any language eventually caves. If it were true that

---

2. Just what "advice" or "questions" one might have while eating at McDonald's is beyond me as well, but this is what it said on one of those paper sheets they give you with the tray.

English were especially "flexible," then there would exist languages that had remained rigidly impermeable to other ones during periods of extensive intercultural exchange–but such languages do not exist.

Russia, for instance, was a highly insular nation for most of its history, viewing "Europe" as an alien and hostile entity. It is no accident that, in their learned writings, they preferred to create Russian versions of Latinate terms rather than use the Latin ones themselves. The nominative case in Russian is the *imenitel'nyj*, meaning "naming" as *nōminare* did in Latin; the accusative case is the *vinitel'nyj*, which has the literal meaning of "to blame" that *accuse* does. *Case* itself is *padež*, from the root "to fall," patterned on the fact that Latin's *cāsus* also meant "fallen."[3] German, more interculturally receptive than Russian but not as much as English, took *Nominativ* and *Akkusativ* but then translated *cāsus* into *Fall*, which has the same meaning as its English cognate. As late as the early 1700s, Peter the Great was advocating that people "write everything in the Russian language, not making use of foreign words and terms" whose use hindered comprehension. Yet Russia still imported a great number of loanwords even in its czarist period, especially from French, in which its elite was often bilingual. And today, as Russia after communism is undergoing a dislocatingly rapid Westernization, the language is being absolutely flooded with English words–the days of studiously translating outsiders' terms into Russian before accepting them is a thing of the past. Quite normal in today's Russian are words like *kserokopirovat'*, from "Xerox," for *Xerox copy*.

Another example is Japan, traditionally one of the most isolated modern cultures in the world. In the past few decades, it has inhaled so many American English words that grandparents find themselves literally unable to comprehend many modern writings

---

3. Technically, Russian in this period did borrow a lot of vocabulary from the "high," mostly liturgical language Old Church Slavonic, used only in writing. Old Church Slavonic was based on Russian's close relative Bulgarian but was used as a written lingua franca in much of the Slavic-speaking world at the time, as Latin was in Western Europe. This was made particularly natural because the sense of Slavic varieties as "different languages" had yet to become as institutionalized, given that the varieties had yet to diverge as much as they have today, and none of them had yet been standardized and developed into vehicles of national literatures.

and signs, heavy with words like *beisuboru* (baseball), *T-shatsu* (T-shirt), *sukii* (ski), *fakkusu* (fax), *bouifurendo* (boyfriend), and one of my favorites, *appurupai* (apple pie). One current as I write is *yan-mama*, from "Yankee Mama"–referring to Japanese women who dye their hair brown or blond to look American.

Languages exerting particularly heavy cultural influence for long periods of time tend to have left particularly large stocks of words in other languages. Because of the spread of Islam, Persian, Turkish, and Urdu (in Pakistan) contain massive percentages of Arabic words penetrating even to basic vocabulary. The Chinese have exerted a similar influence over many Far Eastern cultures. The Chinese occupied Vietnam for more than a thousand years, and as a result about thirty percent of Vietnamese vocabulary is Chinese; as our word for *language* comes from French, Vietnamese's word *ngữ* is from Chinese. As with French and English, most of the Chinese words entered "from on high" and left doublets such as the Chinese-derived *hoả-xa* for *train* in writing and the native Vietnamese *xe lửa* in casual speech.

Japanese was similar long before the American English invasion. The culture has always entailed a paradoxical combination of deep-seated insularity and lively embrace of the externals of other cultures. The use of Chinese symbols as part of the Japanese writing system is paralleled by thousands of Chinese-derived words snaking throughout the language. If you ask the Japanese what the numbers *one* through *four* are, for instance, they will tell you *ichi, ni, san, shi,* which are originally from Chinese. Original Japanese numbers are used as well–*hitotsu, futatsu, mittsu, yottsu*–but these are more restricted in use, given when telling a child's age, for example.

Just as often, however, languages exchange words as the result even of equal-opportunity relations. Of course, many such words enter with cultural tokens new to the language, just as we inherit words like *taco* and *sushi*. But through time, languages in peaceful, utilitarian contact can borrow more deeply than this. In Australia, for example, it is exceedingly difficult to trace any kind of "family tree" among the 260 or so languages spoken there before whites came (since then, as in the United States, almost all aboriginal languages are dead or close to it), because many of the languages have borrowed as much as fifty percent and sometimes more of their vocabularies from other Australian languages. Borrowing runs so wide and deep in Australia partly because there is often a great deal of intermarriage between groups, which ensures that large numbers of people have always been bilingual between pairs of the languages—a prime condition for word exchange. This is furthered by a widespread practice of lexical taboo, under which words that sound similar to the name of a tribe member who died are deliberately replaced by the equivalent word from another language.

Unsurprisingly, then, reconstructing Proto-Australian has proved a much more elusive task than tracing back to Proto-Indo-European was, even though history and common sense require that there must have once been such a single ancestor, spoken by the peoples who first made it to the continent from Southeast Asia. Imagine trying to figure out what the original Proto-Indo-European word for *father* was if you couldn't deduce that the Latin word was *pater*, because French had replaced *père* with *Vater* from Germany next door and Spanish had replaced its *padre* with *pai* from Portuguese; nor could you figure out that French's *Vater* was borrowed from something Germanic, because English had borrowed *pai* from Spanish across the Strait of Gibraltar, whereas the Scandinavians had replaced their *fader* with *isä* from Finnish, and meanwhile over the millennia before you got to the scene, all of these words had been merrily morphing along through sound changes into new words barely recognizable as derived from the original ones.

Dialects Are Still All There Is

We see, then, that languages not only change but mix. We can now finesse the picture a bit by bringing in the dialect part of the equation. Properly, it is often less that language A mixes with language B but that one dialect, or some dialects, of language A mix with one or more dialects of language B. The Normans who conquered England, for example, brought the Norman dialect of French with them, not the one spoken around Paris that was to become the standard. We saw that Parisian *charbon* was *carbon* in the Norman dialect; because of this same difference, our word is today *castle,* from Norman *castel*, rather than *chastel*, the Parisian version (which later developed into *château* and was borrowed by English *again* to refer to the kind of domicile that the word had by then evolved into connoting). Another difference between the Norman and the Parisian dialects was that the former had *w* where the latter had *gu*.

Double-dipping between the two dialects is why we have both *warranty* and *guarantee*, *reward* and *regard*, and *warden* and *guardian*.

At the same time, different dialects of English underwent different borrowing experiences, although the influence of the standard has since gradually spread its particular French acquisitions to the others. Scots English took on some French words that dialects to the south did not; hence *bonny* "beautiful" and *fash* "to bother" (from French *fâcher*). Because the Norse-speaking Viking invaders settled in what became Scotland, Scots also has a stronger Norse imprint than does Standard English. Here is the first sentence of the Prodigal Son passage in Chapter 2: *There wis aince a man hed twa sons; and ae day the yung son said til him, "Faither, gie me the faa-share o your haudin at I hae a richt til."* The *til* for *to*, *gie* for *give*, and *richt* for *right* would all be familiar or reminiscent to the Scandinavian, and things like this render Scots a bit more transparent to a person who happens to know Swedish, Danish, or Norwegian.

Finally, a language can borrow not only from different dialects of a language, but from different historical stages of a language. Japanese first borrowed (Mandarin) Chinese words from Middle (Mandarin) Chinese, then from the standard variety of (Mandarin) Chinese as it had evolved by the T'ang dynasty in the late 600s, and finally from the (Mandarin) variety as it had evolved not long before today. In some cases, the same word was borrowed at two of these stages, leaving variants of the same Chinese word having related but different meanings.

Thus charting where a word comes from in a language is often a matter of tracing not only the language but the dialect, as well as identifying the time slice of the borrowing, because words can differ considerably in shape from region to region and from one era to another, as we see in cases like *castle* versus *château*.

Language Mixture Level Two: Grammar

As I have shown, a language consists not only of its words, but also of its particular sounds and its sentence structures. There is no reason why language mixture should stop with just the words, and indeed it does not. Words are definitely borrowed most easily: Japanese has incorporated hundreds of English words without becoming at all similar to English in regard to its sounds and grammar, for example. However, in situations where large numbers of people are bilingual between two given languages, the two languages often come to resemble each other on the level of sounds and sentence structure as well as the exchange of words–rather like married couples who gradually begin looking like each other over the decades. This happens most readily when literacy in the language is not widespread, such that there is relatively little sense that the standard variety–if there even is one–is The Language, which one flouts to the detriment of one's social legitimacy. For that reason, in today's Big Languages, this kind of contact-induced transformation happened mostly before the past few centuries. Yet its effect played a major part in determining what these languages are like today, as it has in all of the world's languages, most of which, after all, are rarely if ever written and are largely used orally.

The Joys of Yiddish

Yiddish is, in the strictly taxonomic sense, a variety of German. It does not differ from Standard German appreciably more than do Swiss German, Bavarian, and such Austrian German dialects as Viennese, and if it were spoken within Germany by Germans rather than being the language of a culturally distinct group (*Yiddish* means "Jewish," for example), it would simply be considered a German "Mundart" like those other dialects. Yet Yiddish does have many features that are due to its history as a German dialect spoken for centuries by people bilingual in Slavic languages like Polish and Russian. In German dialects, the feminine ending is *-in*: a female tailor is a *Schneiderin*. In Yiddish, this ending is *-ke*, as in *schnajderke*, and this is also a feminine suffix in Slavic: a male American is an *amerikanets* in Russian; a female one is an *amerikanka*. In German, reflexive pronouns change according to person as in English:

ich wasche **mich**	I wash myself
du wäschst **dich**	you wash yourself
er wäscht **sich**	he washes himself

Yiddish, however, just uses the *sich* "he/she/it" pronoun (*zikh* in Yiddish) for all persons, which is exactly what Slavic languages such as Polish do, thus in the literal sense saying *I wash itself, you wash itself, he washes itself*:

Yiddish	**Polish**
ix vash **zikh**	ja myje **sie**
du vashst **zikh**	ty myjesz **sie**
er vasht **zikh**	on myje **sie**

Linguistic Melting Pots

Language mixture has particularly wide-ranging effect when there is extensive bilingualism and multilingualism in assorted languages. In situations like this, several languages will gradually come to share a core of similarities that they did not before, such that each language ends up with traits unheard of in its close relatives that history did not shepherd into this hotbed of multilingualism.

The classic example is in the Balkans. There was formerly a great deal of bilingualism and multilingualism in this area–between Greek, the idiosyncratic Albanian (which, though Indo-European, has a branch all to itself), the three closely related Slavic languages (Serbo-Croatian, Macedonian, and Bulgarian), and Romanian (which despite its speakers' history in communism is a Romance, not Slavic, language). What happened here is perhaps easiest for us to grasp through the language closest to ones most readers will be familiar with, the Romance "outlier" Romanian.

Many language-heads, as I call them, making their way in school or on their own through French, Spanish, Italian, and Portuguese, eventually decide to give Romanian, the "other" major Romance language, a spin, expecting another variation on the basic game plan common to the other Romance languages. Yet Romanian ends up presenting one trait after another that you never encountered in the Romance languages of Western Europe. Perhaps the first thing that jumps out is that the definite article comes after the noun rather than before it. Thus French has ***l'**homme* for "the man," Spanish has ***el** hombre,* Italian has ***l'**uomo*–but in Romanian, it's *om**ul**. Om* is descended from Latin's *homo,* just like *homme, hombre,* and *uomo;* no problem there. But what is the definite article doing over on the "wrong" side?

The reason Romanian is the "twisted sister" in the Romance family is because it was spoken bilingually so often and for so long with other languages in our Balkan group, whose articles already happened to fall after rather than before their nouns (as is true of a great many languages in the world). In Bulgarian, ***the*** *man* is *mŭzhŭt*; in Albanian, it is *burrë-i.* Just as Polish left its one-size-fits-all reflexive in Yiddish, Albanians left their definite article placement in Romanian (the Bulgarians picked it up later).

In other Romance languages, the infinitive, as in *to love,* is just one word–that's one of the first things we learn in French or Spanish, that *I want to eat* is *Yo quiero comer*, not *Yo quiero a comer.* Romanian has a two-word infinitive like ours: *a adresa* "to address," because this cluster of Balkan languages prefers English-style two-word infinitives. Greek has developed one over time (Ancient Greek had the one-word type), and Bulgarian and (some dialects of) Serbo-Croatian have developed one in contrast with other Slavic languages, which also have the one-word infinitive.

Thus Romanian departs from the Romance pattern in large part because of its particular contact history. What is notable about this Balkan situation is that each of the languages in the Sprachbund (German for "language group") is, in the same way, queer in comparison with its close relatives. For example, Bulgarian is unusual among Slavic languages in having articles at all. This is why, for most Slavic speakers, mastering English articles is the very hardest thing in the language: in *Rocky and Bullwinkle,* Boris Badenov's and Natasha's dropping of *a*'s and *the*'s is caricatured but based in the reality of how many Russians tend to speak English if they have not mastered it fully, and it is not uncommon to meet a Slav who speaks English with barely a trace of an accent and with full command of even subtle aspects of the grammar and the raciest idioms but still makes some mistakes with the articles. Among Slavs, Bulgarians have less trouble with this, because they have a definite article, which developed because Romanian, Albanian, and Greek had them.

This group is called the Balkan *Sprachbund.* A Sprachbund is analogous to four college students who share an apartment coming to share a common core of musical, culinary, sartorial, and even social tastes in the course of four years, eventually even settling into the same sleeping patterns. Sprachbunds like this exist worldwide between languages sharing the same mouths.

In India, there is a Sprachbund no larger than a single villag In Kupwar, there are languages from two families spoken. Ur and Marathi are Indo-European languages related to Romar Germanic, and Slavic. Native mostly to the southern regior

India are languages of a smaller family called Dravidian; the Dravidian language we encounter most often among immigrants to America is Tamil [TOM'l], whereas the one spoken in Kupwar is Kannada [CON-a-duh].

In this village, speakers of Urdu and Marathi share space with Kannada speakers. The language divisions are also caste divisions, and for centuries men in the village have switched between all three languages in public interactions, the caste system making it more appropriate to speak to members of other castes in their own language than to imply social parity by communicating in one's own. The result is that the varieties of Urdu, Marathi, and Kannada spoken in this village are now quite distinct from those spoken elsewhere, the reason being that, over time, the three have moved toward each other into a Sprachbund grammar unknown anywhere but this village. Kupwar men can speak any one of the three languages by plugging its words into this ready-made local sentence structure.

For example, although to English speakers a sentence like *Your house big* instead of *Your house is big* sounds "primitive," worldwide there are as many languages if not more that do not use a verb *to be* in such sentences as those that do (and there is no correlation with "civilization" or literary development here–Russian, for example, has *Your house big*). Urdu and Marathi happen to have a *be* verb in such sentences. Kannada, however, of the Dravidian family, is a "Your house big" language, except in Kupwar, where it now uses a *be* verb in sentences like that, because it has been spoken alongside Marathi and Urdu for so long. Here is *Your house is big* in the languages of Kupwar–in all of these languages, the verb comes at the end, by the way:

English:	*your*	*house*	*big*	*is*
Urdu:	tumhara	ghər	bəda	hay
arathi:	tumcə	ghər	mothə	hay
dard Kannada:	nim	mənə	doddu	
r Kannada:	nimd	məni	dwədd	eti[4]

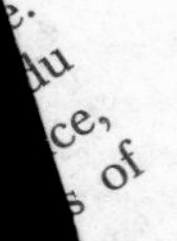

———

speak or know Urdu or Marathi, the reasons these examples
ent from the standard varieties of these languages is because
the Kupwar dialect of both. I have not marked this explic-
oses of streamlining, in order to bring out the point about

Meanwhile, in other ways, Urdu and Marathi have become more like Kannada. In Marathi and Urdu, as in most Indo-European languages, inanimate things are marked for gender just as people and animals are (for example, *el sombrero, la casa*). In Kannada, however, things are easier: *all* inanimate objects and concepts are marked as neuter. In the Kupwar dialect of Marathi, all inanimates are marked neuter just as in Kannada, whereas the local Urdu lumps all of its inanimates into the masculine category, thereby becoming as user friendly as the Kannada system by restricting gender marking of any kind to things biologically female.

Language Mixture Level Three: Intertwined Languages

There is an even further degree of language mixture that linguists have encountered occasionally in various parts of the world. In the examples we have seen so far, despite how many traits a language has taken on from another one, it remains identifiable as the language it was before the contact. The Slavic remnants in Yiddish left a language still clearly descended from German varieties, not one we would think of as a variety of Polish: the basic vocabulary as well as the bulk of the grammar remained intact. Bulgarian has oddities compared with other Slavic languages owing to its contact with Greek and other Balkan languages, but to Slavic language speakers it is still highly similar to their own. But there are cases when two or more languages mix so very thoroughly and deeply that the result is no longer classifiable as either one or the other: instead, the language is a hybrid, just like a mule is neither horse nor donkey.

They Think They're Kidding!

To get an idea of what these languages are like, let's look at an intentionally humorous advertisement that unwittingly illustrates something that happens to languages in reality. In modern Ger many, English is nothing less than a mania, and German advertis ments are so full of English words that at times an American c forget being in a foreign country. On the next page is one of th McDonald's tray flyers, this time from Germany, advertising Egg McMuffin.

McMorning
All American Breakfast
Egg McMuffin™:
About this Frühstücksei lachen ja the chickens.
nur DM 2,95
This is the Preis. Guck at it and think: Oh, very günstig!
The first Weizenbrötchen. So knusprig like the second Weizenbrötchen, which is unten.
The würzige Stück Speck. Gives you a lecker Geschmack and makes funny Geräusche zwischen the Zähne.
And this is from the Hühner: A crazy good Ei.
This ist the Chester-Schmelzkäsezubereitung. Käse is very lecker in the morning.
The second Weizenbrötchen. So knusprig like the first Weizenbrötchen, which is oben. Please do not verwechseln oben and unten!

The headline sentence means "The *chickens* are laughing about this breakfast egg." The word order parallels what the sentence would be in pure German, plugging English words in: *Uber dieses Frühstücksei lachen ja die Hühner.* The next sentences translate as:

> This is the price. Look at it and think: "Oh, very reasonable!"
>
> The first wheat bread roll. As crunchy as the second wheat bread roll, which is underneath.
>
> The savory piece of bacon. Gives you a delicious flavor and makes funny noises between your teeth.[5]
>
> And this is from the chickens: A crazy good egg.
>
> This is the melted Chester cheese food.[6] Cheese is very delicious in the morning.
>
> The second wheat bread roll. As crispy as the first wheat bread roll, which is up top. Please don't switch the top and the bottom!

The modus operandi of this flyer is to sprinkle an underlying German grammar with words from English. For example, "So knusprig like the first Weizenbrötchen" is based on the German *So knusprig wie* . . . , which translates as "so crunchy like" instead of English's *as crunchy as.* The humor comes from the implication that no one would go *this* far in recruiting English words into their German.

Life Imitates Art

And in Germany, they wouldn't. Yet there do exist languages exactly like this. In the 1700s and 1800s, French Canadian trappers often produced children with Native Americans speaking Cr
These children grew up identifying themselves neither as Fr
Canadians nor as Cree but as a mixture of both, and tha
fiercely enough about this status to found communities of thys,
They manifested this bicultural identity not only in thei

5. It does? Cultural difference, I guess.

6. That doesn't sound any more appealing in German, b… asier to imagine making funny noises between your teeth with

but also in a language of their own that neatly mixed French and Cree. In areas along the U.S.–Canadian border, some of their descendants still speak this language, called Michif, today. Just as the German McDonald's flyer plugs English words into German grammar, Michif plugs French words into Cree grammar. More specifically, with few exceptions, the nouns in Michif are from French, whereas the verbs are from Cree. Here is a pure Cree sentence:

Kînipiyiwa	ostêsa	aspin	kâ-oskinîkî-wit.
died	her brother	when	that-young woman-is

Here is a Michif sentence, with the French words in bold type:

Kînipiyiwa	**son frère**	aspin	kâ-**la-petite-fille**-iwit.
died	her brother	when	that-the-little-girl-is

Her brother died when she was a young girl.

The basic grammar is still Cree, in which the verb comes first, and *little girl* is wrapped in a package with *that* and *is*. Seeing or hearing Michif, the outsider might well suppose that these people are just randomly mixing French and Cree or that they are simply compensating for not being able to speak either language well. Neither is in fact the case. If they were randomly mixing the languages, then some nouns would be Cree and some verbs would be French; instead there is a set pattern used by all speakers of the language. Other pieces of the language are specifically assigned to one language or the other, quite regularly from speaker to speaker—*this* and *that* and the number *one* are from Cree, the definite and indefinite articles and the other numbers are French, etc.

Michif is not a fallback strategy for people who could not really
anage their ancestors' languages, nor is it a jolly sort of pig Latin—
a new language altogether. Until recently, Michif speakers were
trilingual in French, Cree, and Michif. Today, they are largely
their Cree, as most Native Americans are losing their origi-
uages, and tend to be English dominant. But these devel-
re only recent.
wa es can melt together like this in an infinite variety of
anc arly common are those in which speakers retain their
new mar but replace the vocabulary *completely* with a
This has happened in the Andes, for example.

Many Indians in Ecuador speak Quechua. Young men from Quechua-speaking villages have often spent long periods of time in the capital, Quito, where they learn Spanish and gain a sense of themselves as having one spiritual foot in their village and another as men of the city. This is an identity that calls for a language half Quechua and half Spanish, and these men created one called Media Lengua ("middle language" in Spanish). In Media Lengua, all of the vocabulary is Spanish, but the grammar is completely Quechua's, down to even subtle details. In Spanish, *I come to ask a favor* is *Vengo para pedir un favor.* In Quechua, it is:

Shuk fabur-da maña-nga-bu shamu-xu-ni.
one favor ask come-ing-I

Quechua, like Latin, has endings that show the function of a noun in the sentence. In Latin, if one sees a dog, one sees not a *canis* but a *canem*, to show that *dog* is the object. In Quechua, the ending *-da* serves the same function. The precise function of the endings after *maña* "to ask" would take us too far from the discussion, but suffice it to say that Quechua is a language that makes great use of endings to chart words' functions. Quechua is also one of many languages in which the object comes first rather than the subject. With these things in mind, here is *I come to ask a favor* in Media Lengua. The Spanish words are in bold:

Unu fabur-ta **pidi**-nga-bu **bini**-xu-ni.
a favor ask come-ing-I

Clearly, this is nothing any Spanish speaker would understand (other than one who also spoke Media Lengua), but then with these Spanish words it would be equally incomprehensible to grandmothers back in the village who only speak Quechua. This is a new language entirely, shoehorning Spanish into a Quechua mold. Notice that, though the Spanish sentence has all five cardinal vowels, *a, e, i, o,* and *u,* the Media Lengua sentence has only *a, i,* and *u.* This is no accident: look at the Quechua sentence and see the same thing. Quechua doesn't have distinct sounds *e* and *o,* and thus Media Lengua speakers render Spanish words with their closest equivalents to these sounds–in other words, with a Quechua accent. Also, note that *favor* in Quechua is *fabur,* from Spanish *favor*–this is a Spanish word borrowed into Quechua that even

Indians back in the village might use. It is the young men who mix the languages beyond the level of words and create something new.

Languages Mating Worldwide

Some linguists have called languages like this "mixed languages," but this is an awkward term because all languages are "mixed" to some degree. More recently, the term "intertwined languages" has been gaining ground, especially as more are discovered. Besides the interest inherent in the languages themselves in regard to their hybrid grammars, the social situations giving birth to them are never dull.

In the mid-1800s, Russian fur trappers produced children with Aleut-speaking Eskimo women on two islands in the Bering Strait and left forever shortly thereafter. The women never spoke Russian well (I can never quite believe that anyone does, the grammar is so complex and riddled with exceptions), nor did the men speak anything but utilitarian Aleut (which was doing pretty well because Aleut is so complicated that it's a wonder its speakers have any room left in their brains for motor coordination, vision, or processing smell and taste). Their children–the world's only Russo-Eskimos–developed a language that uses Russian endings on Aleut verbs a good part of the time. This language is called Mednyj Aleut–"middle Aleut" in Russian.

Gypsies are well known for occupying the margins of the societies they inhabit; meanwhile, as time goes by, they tend to lose the ability to speak the Romani language native to them, a language of India related to Hindi. In England, their linguistic response to their sense of themselves as outsiders is to plug Romani words–still recalled even when speakers no longer know the grammar itself–into English grammar. This language is used as an in-group code to prevent being understood by outsiders. Romani words are in bold:

The **Beng wel**'d and **pen**'d: **Av** with **mandi.**
the Devil came and said come with me

The Devil came and said, "Come with me."

We would predict that, if intertwined languages result from a bicultural identity, Gypsies would create similar languages in other countries they inhabit, and this is indeed the case; languages like this among Gypsies have been reported in Scandinavia, Basque country, Spain, Greece, Armenia, and Yugoslavia.

In South Africa, the languages of writing, high culture, and urbanity are Afrikaans (a close relative to Dutch) and English, both brought in by white colonists. Black South Africans grow up with various indigenous African languages. Zulu-speaking men who have cast their lot with life in Johannesburg have developed a new language in the same way as the young men in Ecuador did. Feeling distinct from their kin and friends back home but obviously not feeling that relocation to the city has transformed them into kindred of ex-prime minister Botha, their new language uses Zulu grammar with vocabulary drawn considerably from Afrikaans and to a lesser extent from English. One of its names is Isicamtho, and the *c* is actually pronounced as a "click" that sounds rather like an English speaker's *tsk* in *tsk tsk*.

Thus intertwined languages exist worldwide when history leads certain groups to feel resolutely split cultural allegiance. There are even cases in which the number of languages is more than two. One of the oddest languages on earth is Wutun, spoken in western China, which, roughly speaking, uses Mongolian grammar with words from (Mandarin) Chinese and Tibetan–but the boundaries are not as definite as with Michif or Media Lengua. There is plenty of Chinese and Tibetan in the grammar, and no little Mongolian in the vocabulary. Spoken in a remote location, by people with little opportunity to travel, and in a country whose linguistics faculties have traditionally had even less interest in language mixture than Western ones do, Wutun is a marvelous chili of a language that unfortunately remains only briefly known.

Finally, like language change and the distance of one speech variety from another one, Linnaeus and Aristotle again fail us when it comes to parsing language mixture, because it is an inherently gradated process. There are no discrete lines on the way from minimal borrowing of words (our use of *tsunami* and *kamikaze* from Japanese) through structural influence of the Romanian type to outright language fusions like Michif and Wutun. There are innumerable cases that fall between the prototypical ones in degree of mixture.

Some languages have, for example, borrowed a larger proportion of words than others–Russian has quite a bit of French-derived vocabulary, but this is a mere drop in the bucket compared with the Norse, French, and Latin inheritance that utterly swamped English. As to structural influence, the Slavic effect on Yiddish was pretty light as these things go; the Balkan influence on Romanian is heavier. There is one dialect of Romanian, though (Macedo-Romanian), where the influence from Slavic in particular is greater than in the standard because of the demographic history of the region. There are Greek speakers in Turkey who have been bilingual with Turkish for centuries. Some dialects of Greek there are suffused with Turkishness in vocabulary, sound, and structure but remain identifiably "Greek"; others are essentially intertwined languages using Greek words with Turkish grammar. Media Lengua is the "tidiest" encounter between Spanish and Quechua, with a neat split between words and everything else. But there are other varieties that are farther along the line toward speaking Spanish itself, with some of the Quechua endings replaced by Spanish ones. In no respect does human language lend itself to binary conceptions and rigid taxonomies.

Shaking Our Fists at the Rain: Language Mixture as an Experience

I didn't like it when I realized that my sister was old enough to date. Of course, I knew it was going to happen some time, and for her sake I wouldn't have wanted it not to ever happen. Yet the idea of men entering and leaving and shaking up her life in the way that I, as a man, was well familiar with doing in other women's lives was distinctly unsettling. It took me a few years to get used to talking about such things with her.

Language mixture is similar in that, although it is a universal process, when we see it happening in our lifetimes we tend to decry it as "contamination." The borrowing of words is particularly constant between languages in contact, being not just common but inevitable. Yet this is the kind of language mixture that gets the most negative attention, under the mistaken impression that rampant borrowing of words from another language is a sign that the process is going to reach its ultimate possible stage, the annihilation of the language altogether. A lot of sincerely worried people in North America and Europe would benefit greatly from realizing that language mixture quite often settles at an intermediate point between *mixture* with language B and *replacement* by language B—furthermore, at a point that remains in all facets identifiable as falling within the bounds of language A.

The Crusade Against "Denglisch"

In Germany today, a business professor has founded the Association for the Preservation of the German Language. Each year this group gives an "award" for "Language Adulterator of the Year" to a public official caught using what the association deems too many English words in the official's statements. In fact, the head of this organization is no zealot; he speaks English very well and has no problem with English words so well established that they are now for all intents and purposes "German" words as well, such as *Sex-Appeal* and *Slip* (that is, the undergarment). He restricts his fire for English words for which there exist German equivalents, suggesting that *downsize* be replaced with *abbauen, Event* with *Ereignis,* and *Highlight* with *Höhepunkt.* "Denglisch"–that is, Deutsch plus English–he feels, is not "real German."

And to be fair, cosmopolitan German speakers, especially in the business and advertising worlds, can be heard using so much English that the McDonald's flyer stops looking so much like a joke. A fashion designer got the 1997 "Language Adulterator of the Year" award for this statement during an interview:

> Mein Leben ist eine *giving-story.* Ich habe verstanden, dass man *contemporary* sein muss, das *future*-denken haben muss. Meine Idee war, die *hand-tailored*-Geschichte mit neuen Technologien zu verbinden. Und für den Erfolg war mein *coordinated concept* entscheidend, die Idee, dass man viele Teile einer *collection* miteinander *combinen* muss.

> My life is a giving story. I have come to understand that you have to be contemporary, be future-oriented in one's thinking. My idea was to combine the hand-tailored thing with new technologies. And what was decisive in making this a success was my coordinated concept, the idea that you combine several pieces from a collection.

But the problem with setting up an organization to resist this is that it is utterly futile. For Germans, as leaders of the world's business and cultural community, to *not* have begun using English to this extent–especially with the speed and density of today's communications technology–would have been as bizarre as exhaling on a cold day and not seeing your breath in the air. The organization's discomfort with English terms replacing suitable German ones seems founded on a certain logic, but the logic flounders from the comparative perspective. Not one of the words longer than four letters in the preceding sentence is an original English word except *English*–but how is this a problem, and how was it ever?

The organization's zeal is tied in part to an understandable sensitivity about the resuscitation of Germany's national pride; its founder, Walter Krämer, has justified his efforts as helping to restore a "missing sense of a right to exist" among Germans. But one wants to ask Krämer: Did the overwhelming of the English vocabulary by French and Latin prevent Great Britain from regaining a sense of its right to exist after the Norman occupation? Even at their most culturally insular historical stage, the Japanese were using Chinese words for concepts as basic as meat and their numbers, and yet the Japanese people have rarely lacked for a sense of legitimacy.

The Association for the Preservation of the German Language and similar organizations, such as the Académie Française's famous vigil against English words "taking over" in French, are laboring under a misconception that, for all intents and purposes, a language is just its words. This is a natural feeling, because words are what we are conscious of–we don't think about the endings and the grammatical rules when we talk. There is a folk conception that learning a language means picking up the word for *hat*, the word for *sleep*, etc.–but we all know that, in actuality, you could memorize thousands of a language's words and still have less ability to communicate than a three-year-old, because how the words are put together is equally central to what "a language" is. English, after all, lost most of its original words. However, because we still use a sound system similar in broad outline to that of our Germanic relatives (it's not hard for us to get a good German accent, and vice versa), as well as a broadly similar grammar (for example, adjectives before the nouns), and retain original English words as our bracing beams and girders (the grammatical words like *will* and *the,* basic words like *boy* and *like*), the language remains fundamentally Germanic–that is, *English.* It could be nothing else–Martians could come down and examine this language and, even though most of the words come from elsewhere, after a while they could tell that all of this was hung on a Germanic foundation.

In this light, there is no danger that these "language adulterators" are leaning toward speaking something "not German." The winner of 1997 is not only using German word order, but in the chinks you can see that she was even using the particularities of German grammar that make German German, as well as difficult for foreigners. *Mein Leben ist eine giving-story,* she says–not *ein giving-story.* She used the feminine *eine* because the word for *story* in German, *Geschichte,* is feminine. In other words, she still had German running "on line" in her head down to its nuts and bolts. In the neurological sense, by all indications the structure of a grammar takes up as much space in the brain as the word collection. This woman has the heart of German intact; she is merely hanging ornaments on it.

"What's All This Spanglish?"

Here in the United States there is a similarly ambivalent relationship to the influence of English on the Spanish spoken by immigrants from Mexico, Cuba, and Puerto Rico. In English, there is so much foreign vocabulary that restricting yourself to Anglo-Saxon–derived words would leave you with the expressive capabilities of, again, a three-year-old. Yet Latinos born here, or who came here at an early age, are often dismissed, or gently teased–which is really a form of the same thing–as speaking "Spanglish" because their dialects of Spanish are bedecked with foreign vocabulary–this time English. The "Spanglish" accusation comes when a Latino says *brecas* for *brakes* instead of the original Spanish *frenos,* or *carpeta* to refer to a rug instead of a folder, as is the case in the original Spanish, or *Voy a manejar mi troca a la marketa* for "I'm going to drive my truck to the market" instead of *Voy a manejar mi camión al mercado.* The sense of Spanglish as "not real Spanish" is reinforced by the fact that relatively un-Anglicized Spanish remains spoken close by in Mexico and the Caribbean, leaving the Americanized dialects to appear "adulterated" in comparison.

Yet what we are faced with here is the development of a new dialect of Spanish, distinguished from Mexican, Cuban, or Puerto Rican Spanish by how much mixture from English has occurred. This dialect is, after all, an inevitable result of Spanish being spoken with English over generations' time in this country. We must ask: If a Spanish speaker (1) speaks English with various people all day long almost every day, (2) encounters English in ninety percent of the written language she comes across, and (3) is bathed daily in

English on television and on the radio, then why *wouldn't* we expect her to begin peppering Spanish with English words–and a *lot* of English words? Predictably, the Internet revolution has seen a minor flood of English terms enter Spanish in America: *clickear, taipear* ("type"), *imailiar* ("to e-mail"), *deletear, chatear.* President José Lopez Portillo's Commission for the Defense of the Spanish Language, founded in 1982 to stem the tide of English entering Mexican Spanish, has had no more effect than the Académie Française or the Association for the Preservation of the German Language, and today even the grand old Academia Mexicana takes a laissez-faire attitude toward the English invasion.

This is the most useful position to take, because this mixture process is normal. Russian immigrants to America pepper their Russian with even more English words than Russians in Russia do, leading to coinages such as my favorite (if I may), *fakapirovat'*, the New York Russian immigrant's new word for "to fuck up."

Thus not only do languages mix, but as often as not, only a subset of the speakers of a language lead lives occasioning such mixing. Thus one variety of the language will be more mixed than another. Results of this happening in the past are mere faceless data to us: recall that there is more Norse admixture in Scots than in more southerly dialects of English or that one Romanian dialect has more Slavic in it than the others. But when we see a variety of Spanish with a heavy English overlay developing under our noses, we see it derided as "not real Spanish." Obviously, we are being misled by how slow and gradual language change is–but in the end, if it's okay for Macedo-Romanian and Scots, then it's okay for the Latino immigrants in California, Florida, and New York.

The Quest for an "Vnmangeled Tung"

This discomfort with language mixture is by no means only a modern ailment: it would be hasty to attribute it to Germany's particular history or the specific interethnic tensions in the United States. Certainly, the foreign origin of most of English's word stock makes no difference to us today, but this is partly because the words are so ingrained that we do not think of them as foreign. There were people in days of yore, however, when the foreign borrowings in English still felt "other," who considered the preponderance of foreign words in English an intrusion. There is indeed a certain beauty in the uniformity of an idealized English using only Anglo-Saxon roots–symmetry and consistency are pleasing in all forms. Seamus Heaney's translation of *Beowulf,* with its newly created compounds like "frothing wave-vat" for *ocean* mimicking Old English poetry's propensity for coining compounds from basic English words, gives a sense of what English would be like if we had to use the root stock to form "the big words" instead of just using ready-made French and Latin ones.[7]

But by the time the authors in question started complaining, the die was cast, and their comments today can only strike us as a bit forlorn. Man of letters John Cheke instructed in 1561 that "Our own tung shold be written cleane and pure, vnmixt and vnmangeled with borrowing of other tunges," following this with *mooned* substituting for *lunatic,* and so on. Well, there'd be a mashed-potatoes kind of comfort if the tung had stayed vnmixt, but there was nothing John Cheke could have done about it by Shakespeare's time, which is especially clear given that both *pure* and *mangled* come from French!

Language mixture, then, is universal and inevitable. It often looks like the dissolution of the language when it happens, but once it happens, like people who talk on cell phones while driving and the musical scores of Andrew Lloyd Webber, it's there whether you like it or not, and it's going to keep on coming.[8]

Paradigm Shifts: Language Mixture as the Rule

Among linguists, it has always been known that languages regularly exchange words, but mixture of structural patterns was generally treated as a subsidiary phenomenon, a kind of static amid an operation dominated by good old-fashioned language-internal change. The identification of aspects of mixture in languages was subject to an "innocent until proven guilty" frame of reference, the sentiment

7. Technically speaking, though, even *froth* is Scandinavian!

8. I sincerely hope it remains technically impossible to use cell phones on airplanes–can you imagine what that's going to be like? America has already been saddled with the McMusicals of Andrew Lloyd Webber's American descendant Frank Wildhorn, whose *Jekyll and Hyde, The Scarlet Pimpernel,* and *The Civil War* manage the feat of precisely rendering dishwater in musical language.

being to keep mixture-oriented analyses corralled enough to keep "regular" change front and center. This was rather like the feeling some people have about food mixing on their plates–if one must, okay, but always better not to.

This stemmed from a general impression that languages were relatively resistant to sharing grammar, as opposed to mere words. And it is true that words are shared most easily. Yet as a large body of studies documenting language mixture has accumulated–or, more precisely, as linguists have realized the sheer weight of isolated studies of this kind that were done in the past hundred years but rarely engaged as a discrete body of knowledge by most linguists–it has gradually become clear that languages share structure much more readily than formerly supposed, and there is nothing "subsidiary" about the process. The pronouncement of a grand old German linguist in 1871 that "There are no mixed languages" (it sounds even more authoritative in German: *Es gibt keine Mischsprachen*) can no longer stand.

The Frustrations of Australia

Recent work suggests that even across entire continents, where languages and their dialects are spoken contiguously over millennia, migration and contact lead to a constant simmering of a kind of language stew, where after thousands of years all of the languages will be replete with structural traits borrowed from one another through bilingualisms and multilingualisms, most of them now lost to history. This has always been clear in Australia. The hundreds of languages spoken there are ultimately descended from a single original one, brought to the continent about 40,000 years ago by people migrating across Indonesian islands from Southeast Asia (where there still exist isolated pockets of people physically resembling Australian Aborigines). The languages that evolved from this original source are today a marvelous taxonomic bramble. This is partly because of the word taboos, but also because the widespread marrying out of one's group leads to widespread bilingualism, which has in turn has brought many contiguous languages closer together in structure as well as vocabulary. Many is the language in Australia (and New Guinea) where its relationship to its closest kin is readily apparent only through a subset of its words, because it has gradually inherited much or even most of the grammatical game plan from languages of a neighboring subfamily–having all suffixes instead of all prefixes or even inheriting the other languages' very prefixes or suffixes themselves, for example.

Australia, however, has often been seen as "peculiar" in this respect, when in grand view it is really just a rather extreme manifestation of an ordinary situation. This becomes clear when we pull the camera back to the the languages of Europe in general, which on the surface give the appearance of being largely independent systems, neatly enshrined in their own literatures, cultures, and Berlitz books. Surveying this situation also serves as a bird's-eye view of the aspects of language that linguistic analysis often focuses on, many not taught as "grammar" in school and thus not apparent to us as we use our language or others throughout our lives.

The Origins of "European" Grammar

Most of the languages of Europe belong to the Indo-European family. On the other hand, there are other language families in Europe: Estonian, Finnish, and Hungarian are from the quite different Uralic family; and there are many Indo-European languages spoken in Asia, such as Persian and its relatives and the Indo-Aryan languages like Hindi and Marathi.

Given that so many of the languages in Europe are descendants of a single ancestor, we would naturally expect many of them to have common features. What is interesting is that there are many features common to European languages that were *not* part of Proto-Indo-European grammar–on the contrary, in historical documents of the various languages, these features at first do not exist and then only gradually appear over time. What's more, many of these features developed in the Uralic languages as well, where there was no family-descent relationship at all with Proto-Indo-European. Finally, often the very same features have *not* developed in the Indo-European languages spoken outside of Europe in Iran and India.

In other words, languages spoken on the European continent appear to constitute a grand Sprachbund, whose commonalities have arisen through geographical proximity rather than family relationships. What most likely happened is that, just before the Middle Ages, the "Great Migrations" of various European peoples across continental Europe led to a great deal of bilingualism and multilingualism, the result of which was that the languages of Europe came to have in common a core of traits lent by the various languages to one another. Thus the Romance, Slavic, and Germanic languages, Greek, Albanian, Finnish, and Hungarian display a grammatical kinship that they would not if they had not been in such close contact during a period of heavy bi- and multilingualism.

A good illustration of this comes from a feature that English itself happens to lack. The verb *change into,* for instance, can be used in two ways–to refer to something being changed, as in *I am changing the prince into a swan,* or to refer to something changing by itself, as in *The prince is changing into a swan.* In a great many languages, either one kind of "changing into" or the other one must be distinguished with some kind of marker. In French, for instance, *I am changing the prince into a swan* is *Je transforme le prince en cygne.* However, if the prince just sits there and changes by himself, then in French one must say that he is changing *himself* into a swan: *Le prince* ***se*** *transforme en cygne*, not the incorrect *Le prince transforme en cygne*, which to the French person sparks the question, "Well, *what* did he turn into a swan?" This is found in most European languages and gives us English speakers trouble–it was one of the last things I managed to start "feeling" naturally in my sense of these languages: German would have *Der Prinz wandelt* **sich** *in einen Schwann;* Russian, with the reflexive pronoun *-sja* would have *Knjaz' prevrašajet****sja*** *v lebed'*.

Beyond Europe, however, as often as not, a language puts the marker not on the "self-generated" version but on the one where someone exerts the action upon another. In Maori, in New Zealand, if someone changes into a swan by themselves, then they *rerekee*–no marker like the French *se* or German *sich*, just the bare verb. But if you change the person into a swan yourself–in other words, if you *make* that person change into a swan–then the verb takes a prefix and you ***whaka****rerekee* the person into a swan. Thus French is what we can call a "changes itself" language–it's when something is doing something by itself that the verb gets an extra bit of stuff. But Maori is a "make it change" language–if you do something by yourself, the verb is bare, whereas it's if you *make* it happen that you add a bit of stuff to the verb.[9]

---

9. Yes, I am concocting these terms for our purposes. If you really want to know, I am referring to something many linguists would call degrees of *anticausative prominence.*

Now, if European languages tended to be "changes itself" languages simply because Proto-Indo-European had been one, then the prevalence of this feature in Europe would be unremarkable. Yet in fact, to the best of our knowledge, Proto-Indo-European was a "make it change" language.[10] The "changes itself" feature can be seen developing in its various European descendants only *over time.* Furthermore, even Finnish and Hungarian verbs are "changes itself" types about half the time–whereas crucially, in the Indo-European languages of India, it's marking "make it change" rather than "changes itself" that is the order of the day. In other words, it seems that a language situated in Europe tended to come out as a "changes itself" one regardless of whether its ancestor was.

This means that a core of languages spoken in the geographical entity of Europe all drifted in a similar direction, independently of what protolanguage they evolved from. This is one of several indications that, although Sprachbunds can be as small as a village like Kupwar or extend to a compact region like the Balkans, they can also in grand view encompass an entire continent.

Another example is how languages mark subjects when the topic has to do with experience rather than action. When we learn Spanish, one construction we have to pause to wrap our heads around is that *I like books* is *Me gustan los libros* "the books please me" rather than what we would "expect," "Yo gusto los libros." In Spanish, you say not that you like something but that it "pleases to you," and thus "to me please the books." This is not a mere random quirk–it follows from a distinction in types of object that English just happens not to mark explicitly very much.

We get a handle on the distinction in question with the *gustar* case: technically speaking, *books* in *I like books* is an "object." But *book* is an object in a different way in *I like books* than in, say, *I read books.* To like a book is not really to do anything to it–in contrast with reading a book, you can like a book without touching it; if you say you like a book, then you are actually describing a sentiment, not an action. It is not accidental that Spanish has the "to me" construction with *like*, a verb describing a kind of experience, and not with, say, *drive.*

---

10. Proto-Indo-European marked this by changing a vowel inside the verb, and English retains remnants of this in the distinction between verbs like *rise* and *raise* (that is, making something rise) and *sit* and *seat* (that is, making someone sit).

Yet things like this are the exception rather than the rule in Spanish–usually we can get away with ignoring this subtle distinction of types of object when we learn or speak Spanish. This is largely the case with European languages in general. Some European languages overtly observe the distinction somewhat more than others: in Russian, one does not say "I am cold" but "it is cold to me" (*Mne xolodno*). But in Russian it's only a tendency–when a Chekhov character exclaims that he is happy (inevitably just before someone goes off and shoots himself), he says *Ja sčastliv* "I am happy," not "It's happying to me." German favors this distinction to about the same extent (I remember a guy talking about how, after a night of drinking American beer, *Es stinkt mir im Kopf*–"It stinks to me in the head").

In many languages outside of Europe, however, it is very much the usual case that, if a sentence has to do less with exerting an action upon something than with simply having some kind of experience related to it, then you say that it happened to you rather than that you did it, even when describing actions that feel pretty "active" and "object-izing" to an English speaker. Thus you say not that *I saw a horse* but "To-me saw a horse" or, to stretch it a bit, "I've undergone a horse-sighting." This does make sense if you think about it–the horse suffers no effects from your laying your eyes on it; it's you who has various thoughts and impressions, unlike if you kick a ball or sell the horse. Languages spoken in the Caucasus area, such as Georgian and Chechen, are of this type, such that this way of rendering thought is native to, say, Eduard Shevardnadze.

The "dorm room" resemblance of European languages reveals itself in that, though they all tend to go relatively light on "to me" verbs, this is once again inexplicable by mere common inheritance from protolanguages. Proto-Indo-European was by all indications a "to me"-loving language, and we can see the switch to the English-style distaste for it emerging over time in historical documents of, for example, the Germanic languages. Meanwhile, the Uralic clan's Hungarian "disprefers" the "to me" construction just like the Indo-European languages, despite not being descended from Proto-Indo-European; Hungarian's relative Finnish has a bit more of it but only as a tendency. But then crucially, in the meantime, the "to me" construction is absolutely central to basic expression in the Indo-European languages of India; the faintest whiff of an experience in contrast with an action will set off the "to me" alarm. I can't resist giving a quick example. In Hindi, if you deliberately meet with Apu, then the sentence has an "English" feel to it:

Mẽ Apu se mila tha.
I Apu with meet did

I met Apu.

But if in saying *I met Apu,* what you mean is that you ran into Apu by chance in the "Oh, by the way, guess who I ran into today?" vein, then this was less an action than something that happened to you–an experience. And thus you must use a "to me" construction:

Muj-he Apu mila tha.
to-me Apu meet did

I met Apu.

Several other features in the languages of Europe pattern similarly.

One final clincher showing that the European languages have "bled into" one another structurally comes from the Celtic group of Indo-European. Since the Romans, Vikings, and Goths occupied much of western Europe, these languages, such as Irish Gaelic and Welsh, have been pushed mainly to the western parts of the British Isles, with the exception of Breton in France, remnant of a long-ago migration of Celts across the Channel. Isolated geographically and culturally marginalized, Celtic speakers did not participate in the "Great Migrations," and it is almost certainly for this reason that they only fitfully follow the "European" patterns despite their Indo-European origin. Irish, for example, does not feel "family" to the English-speaking learner in the way that French, German, Russian, or even Greek or Lithuanian would. Celtic languages form an unusual group apart compared with any other Indo-European languages in Europe–their verbs come first in the sentence, sound changes have often transformed even basic words beyond recognizability to any Indo-European speaker who is not a linguist, and there are any number of other fascinating grammatical peculiarities that lend an almost cultlike air to the community of linguists specializing in them. *He's making music,* for example, in Irish is *Tá sé indhiaidh ceól a dheanamh*, literally "be he after music a'making." Celtic languages have marched to the beat of their own drummer to this extent in large part because they have not been constrained by contact with the other languages into honing to certain patterns.

Linguistic Macramé: Symbiosis and Language Change

I have analogized language mixture to the mating of a horse and a mule, but one problem here is the implication that language mixture is rather exceptional. Also, because mules are sterile, another problem is the possible impression that hybridicity in language results in speech varieties somehow "broken."

Modern developments in biology suggest a richer and more appropriate analogy. It is well known that certain life forms live parasitically on others: the bacteria in our intestines, for example, or the clown fish that live among sea anemones' tentacles, sharing the anemones' food while returning the favor by keeping the anemones' tentacles clean. Lynn Margulis and other biologists have called attention to the fact that this communal living, called symbiosis, is less anomalous than rife among the world's creatures, being central to the existence of life itself. Plants derive crucial nutrients from the fungi in their roots that process nitrogen for them; cows could not digest their food without the bacteria filling their stomachs; and even the organelles within cells, such as mitochondria in animals and chloroplasts in plants, began as independent bacteria–all of today's cells are symbiotic colonies of what began as separate strains of bacteria. Thus life itself as we know it is, in large part, thriving, evolving variations of symbiotic, rather than individual, life forms.

Another analogue between language and biology is how viruses insert parts of their DNA into that of other organisms, the latter often then carrying these genetic instructions forever more. Baboons and cats, for example, have in common a piece of DNA unknown in other mammals and therefore impossible to trace to the Ur-mammal that gave rise to both of them; the DNA fragment was possibly transmitted into both by a mosquito in the distant past.

These discoveries show that not only the family-tree model of evolution but the "bush" model corrective argued by Stephen Jay Gould and others are idealizations. As Margulis puts it:

> In reality the tree of life often grows in on itself. Species come together, fuse, and make new beings, who start again. Biologists call

> the coming together of branches–whether blood vessels, roots, or fungal threads–anastomosis. . . . Anastomosis, although less frequent, is as important as branching. Symbiosis, like sex, brings previously evolved beings together into new partnerships.

Richard Lewontin similarly observes that the family-tree model ought to be replaced by that of "an elaborate bit of macramé."

So it is with language, not only between languages in one contained area like Kupwar or even between several languages occupying a larger region such as the Balkans, but sometimes between languages by the continent. All of the world's languages as they evolve carry along words, and often even structures, from other languages, such that some languages are even fifty-fifty hybrids of both parents. In broad view, the world's languages comprise tens of thousands of dialects harboring evidence of symbiotic matings in the past.

Take Romanian one more time: within the language itself, Slavic words have ousted many Romance roots–*iubi* for "to love" is cognate to Slavic roots like Russian's *ljubit'* (minus the initial *l*). More generally, Romanian has inherited features from the languages spoken close by, such that it and the other languages in the Balkan Sprachbund carry traits symptomatic of linguistic symbiosis. Some scholars have proposed that effects from the Balkan Sprachbund can be identified even beyond these six languages, and in this light, most of the languages of Europe can be shown to have cross-pollinated one another with a basic grammatical template. Thus a Romanian spoken by a population that had settled in an isolated mountain valley and had little contact with outsiders would be a vastly different language: the soul of modern Romanian emerged through a massive shipment of Slavic word borrowings and a concomitant pull toward features in languages surrounding it; meanwhile, as a European language, the evolutionary choices it made were also contingent on the "consensus" among that continent's languages forged by population migrations.

Margulis describes anastomosis as "branches forming nets," and this analogy is so useful that we can now replace the provisional one of a flowering bush that opened the chapter.

But what about a branch that has its growth interrupted by being cut through or singed but grows back again?

4

Some Languages Are Crushed to Powder but Rise Again as New Ones

A Polish woman who works at a café near my office has taught me just enough Polish to be able to greet her, order things, and say thank you. It's a little running joke between us. People standing behind me in line might be fooled into thinking I "speak" Polish. But I most certainly do not. I cannot conjugate verbs; I know none of the cases; I don't know how to say *impossible, me either, smells like* . . . or much of anything I would need to express thoughts beyond the specific context of my asking her for beverages. And even if I hung out with this woman for a day and learned a hundred or so words for things like *car, walk,* and *yesterday,* I still would not be speaking Polish. I would be speaking a Polish version of the English of Tonto, the Indian on *The Lone Ranger.*

For me, this is just a little game I play once a week or so because, for an English speaker, it is rare to meet anyone who does not speak English. Even if a person does not have full mastery of English, almost anyone we meet can at least fake it–in other words, talk like Tonto to us. Few Westerners ever have much reason to speak a language Tonto-style for any length of time, because almost anyone they interact with either speaks their language to at least some degree or speaks another language that they both know (such as, all too often these days, English).

This, however, is a by-product of the fact that we Westerners live in geopolitically dominant nations; we have long imposed our languages on others elsewhere, and our economic advantages give millions of people an incentive to master our languages on immigrating to our countries or even if staying in their own. For Europeans and Americans, whoever one is going to communicate with in a sustained way speaks either one's native language or a language that one has had an opportunity to learn at least fairly well. If not, there are interpreters.[1]

In "The Real World," however, different languages are often crammed into smaller spaces, no single language having been spread at the expense of the others across great expanses by colonization and imperialism. Papua New Guinea, just a little bigger than Montana, hosts about eight hundred languages; Nigeria could be contained within Texas, Oklahoma, and Louisiana and is home to about two hundred fifty. These are extremes, but even less pronounced cases contrast sharply with the typical Western experience. Part of the reason the Polish woman at the café is so tickled that I even bother to try to speak Polish is that Poles must cope with the fact that, as soon as they leave their country, another language will be spoken. In Côte d'Ivoire, roughly the same size as Poland, dozens are spoken–it is quite normal to Ivoiriens in the capital, Abidjan, to have never heard of the language a person they know spoke as a child in his home village. This means that learning new languages is more central to the warp and woof of the existence of a great many people in the world than it is for, particularly, an English speaker–often stepping beyond one's village or even shopping at the local marketplace means grappling with a brand-new tongue.

Just Enough to Get By: Pidgins

Such contexts condition a more fluid conception of "speaking" a language than does the Western one. Most languages are not written, and thus one acquires them not to have access to a literature or to "learn about the culture," but for the utilitarian purpose of interacting with their speakers for reasons tied to basic sustenance through the exchange of goods and services. This means that one's

1. In the 1980s, a friend mentioned seeing a video clip where Ronald Reagan and François Mitterrand were sitting in a back seat "talking," with no interpreter in sight. But how were they communicating? You could just kind of tell that Mitterrand did not speak English, and Reagan sure didn't speak French.

aims are often served through what we would regard as halting ability, what we would call "speaking a little X."

This hands-on approach to language acquisition is reinforced by the fact that, in most places in the world, there are no materials available for any but a few languages to be taught formally through books and instruction, such that one usually learns other languages "live" through imitation and practice, not through written drills of sentences like *My cousin's friend is wearing a shirt*, studiously corrected by a teacher, or through listening to tapes purporting to depict speakers in their "cultural context" (which in practice generally means having four characters attend some indigenous event, mention how nice it is, and then proceed to converse at length about the fact that their cousins' friends are wearing shirts).

In other words, life in many parts of the world conditions a need to communicate on a regular basis in languages one can only express basic thoughts and needs in—with this being an ordinary and accepted part of life, just as many of us can cook up a stir-fry with no pretense of having mastered the body of Chinese cuisine.

"Me in Yours": Summer on the Fjords

This happens particularly often in trade contexts. An interesting example was in the 1800s, when Russians would bring timber to trade on the fjords of Norway every summer. The Russians had not been instructed in Norwegian before their trips, nor did the Norwegians have any reason to put themselves through learning Russian. Yet they all had to communicate somehow, and sign language won't do for a whole summer.

The way they did this was by learning bits and pieces of each other's language, stirring the two languages up into a makeshift lingo that served nicely as a vehicle for them to trade in with some social interaction on the side. *Sobaku på moja skib*, one of the traders would have said for *There's a dog on my ship*. *Sobaku* is from Russian *sobaka* "dog." *Sobaku* is the accusative-marked form in Russian, but the Norwegians did not learn the Russian cases, because they were not necessary to the very basic communication needed. Thus the traders used *sobaku* no matter how the word was used in the sentence. *På* can translate roughly as "on" in both Norwegian and Russian and thus represented a compromise between the two languages that all speakers would readily understand. *Moja* is Russian for *my* (feminine form, but used across the board in the lingo) but would be relatively easy for a Norwegian to remember because Norwegian's *my* is *min*. *Skib* is from Norwegian's word for *ship*.

Linguists call this Russenorsk, but its speakers called it *Moja på tvoja*, which meant roughly "Me in yours"; Russenorsk was a way for Norwegians to speak just enough Russian and Russians to speak just enough Norwegian to get by.

Unlike nonstandard dialects or mixed languages of various kinds, Russenorsk was not "a language." There were only a few hundred words and very little of anything we would call grammar. To be sure, there were some loosely established conventions: Russenorsk was not just each man tossing off whatever came to his mind: for example, *på* as the all-purpose preposition was a convention–there are many other words in Russian and Norwegian that could have served just as well. But this contrasted with the hundreds of rules that any full language is built on, requiring thick tomes to outline or teach. For example, learning Spanish entails mastering a casserole of sets of endings, with six for person and number for each, to situate an event in time–preterite, future perfect, conditional, subjunctive, etc. Russenorsk, on the contrary, had no such machinery except to tack on the word for *yesterday* or *tomorrow* if absolutely necessary – no *I had seen*, no *I would have known*, no *we are rowing* versus *we row*, no way to specify *the dog* rather than *a dog*, no way to distinguish *although he went away* from *because he went away*. Russenorsk was good for barter, for brief, concrete observations, and, apparently, for drinking. To say something like *A little hint is all we need to nudge him into considering a change of heart* would have been impossible–but then who needed to say such things very often to the foreign traders? After all, it was just for the summer.

This kind of rudimentary but functional way of communicating is called a *pidgin*, and pidgins have always formed throughout the world when people needed to use a language on a regular basis without having the need or motivation to acquire it fully.

As pidgins go, Russenorsk was "equal opportunity," where the power relation between the two groups was fifty-fifty, such that the traders met halfway, with about half of the words coming from Russian and the other half from Norwegian. Usually, however, sociohistorical realities are such that one group has its foot on the other's neck, and the subordinate group is compelled to make do as best it can with the dominant group's language, rather than the two groups mutually accommodating to each other's. As such, in most pidgins, the bulk of the vocabulary is drawn from the dominant group's language.

Other Pidgins

This was the case with, for example, Native Americans when whites first occupied North America. Today, most Native Americans in the United States speak English, and most of the indigenous languages are gradually becoming extinct (only three hundred thousand people in the United States speak any of the one hundred or so that survive, most of them fast on their way to extinction). But in the 1600s, Europeans encountered an America where no European language had yet eaten up the indigenous languages, such that the distribution of languages across the continent was normal–that is, typical of areas before they are colonized by a large modern power. In such areas, relatively compact groups speaking separate languages coat the land in a patchwork-quilt pattern; although some groups inevitably conquer or exterminate others, this is not widespread enough to alter the basic picture. In America, then, as many as three hundred Native American languages were still spoken by thriving cultures, all having developed as offshoots of one, two, or three languages first brought to the continent.[2] Until the 1800s, typical Native Americans, if they learned English at all, learned it only as a medium for basic communication with alien invaders (and thus Pocahontas would have spoken a distinctly nonnative English with a thick accent, not in a voice sounding like Marlo Thomas).

This was a prototypical situation for the birth of a pidgin English. When we see Indians depicted in old movies as speaking a fragmentary English along the lines of "Over hill lies big heap buffalo," our modern sensibility is to dismiss this as a stereotype implying that Indians were incapable of learning English, and certainly Hollywood, as was and is its wont, took this ball and ran with it to the point of caricature. Yet this stereotype, like most of them, had its basis in a kernel of truth. Through to the 1800s, before Native

2. The traditional view that these languages all trace to a single migration over the Bering Strait from Asia has become increasingly less unequivocal as evidence mounts that some humans may have arrived from Southeast Asia and other East Asian locations.

Americans were deprived of their original languages through assimilation and often physical punishment in American-run schools, many of them did speak a pidgin English. This was not because of linguistic sloth but because this was all that was necessary for someone still rooted in his native culture and only needing English for trade and brief interactions. Native Americans across the country really did often say *heap* for *a lot* and *squaw* for *woman.* These were some of a small number of regularities that one observed when speaking this pidgin.

Even with its limited resources, with a bit of creativity one could get a great deal of life's basic work done in this pidgin English. Here is my favorite passage, from a woman dissing a white suitor:

> You silly. You weak. You baby-hands. No catch horse. No kill buffalo. No good but for sit still–read book.

As we see, English was the main source of the words in this pidgin. *Squaw* was from the Narragansett language but was an exception amid a vastly predominant contribution from the language of the people who were gradually acquiring control over the Native Americans' lives.

There is another example of a stereotype based on an actual pidgin in the "forbidden" episode of the Belgian *Tintin* series, about the worldwide adventures of a bizarrely anodyne fellow and his little dog. Withdrawn from circulation in America is *Tintin au Congo,* originally published in 1946, where Tintin encounters "natives" drawn in a thick-lipped minstrel style now unthinkably backward and insulting in our modern American context. What makes it worse is that all the blacks speak a "broken" French, which makes them appear to be imbeciles, especially considering that a great many blacks in countries colonized by France and Belgium speak marvelous French or at least a French vastly more fluent than what the "darkies" of all social strata in *Tintin au Congo* speak.

Yet, ironically, linguists have analyzed the *Tintin au Congo* text and found that the blacks' particular kind of French parallels unexpectedly closely some particular characteristics of the actual pidgin French that many Africans new to the language have long spoken. Somewhere along the line, author Hergé had apparently been exposed to French West African Pidgin French.

One black praising a train says *Ça y en a belle locomotive.* In French itself, this would be a barely translatable sequence meaning something like "That's a nice train of it"; *y en a* is by no means an obvious or inevitable way to render *to be* in French at any level of familiarity with the language. This is why it is so interesting that Hergé has the character say this and that the black characters pepper their speech with *y en a* throughout the episode, because this really did happen to be a construction that became conventionalized, by chance, as *to be* by Africans using French "on the fly" as a pidgin.

Other features used by the blacks are also typical of real pidgin languages. Just as a Native American Pidgin English speaker would use one form for the first person singular, *me*, instead of alternating between *I* and *me* as we do in actual English, the blacks in the *Tintin* episode use *moi* in place of *je* and *me* in French. Where a French speaker would say *Je vais me salir* for "I'll get myself dirty," a black man in the episode says *Moi va salir moi.* When a driver says *Li machine li plus marcher* for "The engine doesn't work anymore," he does not conjugate the verb, which would be *marche* in French. This is also typical of a pidgin, just as Russenorsk verbs were the same in all persons and numbers despite both Norwegian's and especially Russian's various conjugational endings.

When a Pidgin Becomes "What I Speak": Creoles

The crucial thing about pidgins is that their speakers are using them mostly as transitory tools for passing exchanges. The Russians and Norwegians up in the fjords in the 1800s, the Native Americans speaking a pidgin English, and the black pidgin French speakers in the *Tintin* episode all went home and spoke the native languages they lived in, just as if we spend a Sunday working up a Spanish paella, the lunch we make to bring to work the next day (if it's not leftover paella) is much more likely to be a turkey sandwich than an assortment of tapas. As such, in themselves, pidgins are incidental to living, natively spoken languages. They are mere half-languages used on the side, a kind of spilling of linguistic seed.

Sometimes, however, social history has thrust pidgins into a more influential role in the eternal developments of the thousands of descendants of the first language. In certain unusual situations, humans have been forced to use pidgins not as make-do lingos but as primary languages in their lives, sometimes the only one. In situations like this, humans display a primal disposition to express themselves in a precise and flexible manner. Thus if deprived of the opportunity to use the languages they were born with, they refashion pidgins into brand-new languages. Because as a rule any language spoken on earth traces back to unbroken development from a former full language (or languages), when we see pidgins transformed into creoles we come closest to witnessing the birth of a human language.

Language Born Anew Down Under

In the late 1700s, during their colonization of Australia, the English encountered Aborigines along the eastern coast of the continent speaking various languages. The natural result of trading with them and eventually yoking them into service of various kinds was that the Aborigines developed a pidgin English for transitory use with whites, which whites often spoke with them as well, because this ensured better communication than did using English itself. In the next century, the English expanded their domain into the Oceanic islands east of Australia, first through the whaling trade, then in collecting sandalwood and sea cucumbers. All of these endeavors required cheap manual labor, and geographical practicality meant recruiting men from various South Sea Islands. Gradually, a loosely definable set of linguistic norms for speaking "English," seeded by the Australian Aborigines and brought along for use with the Melanesians, became established among indigenes in the South Seas.

An early example of this pidgin was recorded by Richard Henry Dana in 1835:

> No! We all 'e same a' you! Suppose one got money, all got money. You–suppose one got money–lock him up in chest. No good! Kanaka all 'e same 'a one.

(*Kanaka* is a word for "islander.") Two more examples are these from the 1850s:

> Me think missionary stop board that ship.

> Suppose missionary stop here, by and by he speak, "Very good, all Tanna man make a work." You see that no good . . . Tanna man no save work.

From the 1860s:

> You plenty lie. You 'fraid me se-teal. Me no se-teal, me come worship. What for you look me se-teal?

Like Russenorsk, Native American pidgin English, and the *Tintin* pidgin, this South Seas pidgin was not *utter* language soup. *Stop* had the specific meaning of "to be" when referring to location. *By and by* was the closest thing to a future marker in the pidgin. *Save* is cognate to the "savvy" familiar from old cowboy movies. In the late 1400s, the Portuguese were the first European power to explore the west coast of Africa, and a pidgin Portuguese developed there. As other European powers began sailing the coast, pidgins based on their languages developed as well, but all of them inherited some bits and pieces from the original Portuguese pidgin. *Savvy*, from Portuguese's *sabe* "knows," was one of those pieces and was later carried by sailors into pidgins used in North America with Native Americans as well as those used in the South Seas.

Now if the story stopped here, with Australian Aborigines and Pacific Islanders simply using the language to haggle with and tell off white people while working under them briefly, then we would have just another passing pidgin, which would now most likely exist mainly in old travel narratives, as many pidgins do. But history thrust this pidgin into a special role.

In the 1800s, the English established plantations in Queensland on the eastern coast of Australia and in Fiji, the Germans meanwhile establishing plantations on Samoa. In earlier centuries, the tradition would have been to bring Africans to work the plantations as was the case in the Caribbean, but by the mid-1800s, the outlawing of the slave trade by the British had cut off this source, and thus Melanesians of various islands and men from nearby Papua New Guinea were signed up for short-term contracts (often under duplicitous conditions that rendered their work conditions in effect a new kind of slavery). The pidgin English that began in Australia and had been passed on to the sandalwood and sea-cucumber trades was used on these plantations as well, often transmitted by

Melanesian foremen who had worked in the previous trades. The Germans even accommodated to this new pidgin in Samoa, allowing it to take root there in place of the German pidgin that would have arisen and flourished otherwise.

Because the islanders spoke several different languages, the pidgin, as a simple and easily learned little system, was a natural and easy solution to the communication problem. What this meant, however, was that the pidgin was being used not just "on the fly" but all day every day over decades' time by thousands of people.

Nor did the language die with the plantations. In Papua New Guinea, for instance, as I noted, about eight hundred languages are spoken, often changing from village to village. Laborers returning to Papua from plantations in Queensland, Samoa, or Fiji found that the pidgin was useful in communicating with people from other villages, obviating the need to grapple with the evolved complexities of the actual native languages themselves. First German and then English missionaries in Papua New Guinea found the pidgin similarly useful and thus encouraged its use and spread. Other laborers returning to their native islands, especially the Solomons and Vanuatu (formerly known as the New Hebrides), often found similarly that the pidgin was a convenient vehicle for fellowship between them and other former laborers with different mother tongues, and they often even nurtured the pidgin as a badge of having come through the plantation experience.

Crucially, for people who are speaking something all the time, a language where the likes of "You baby-hands" and "Moi va salir moi" are about all one can get across just will not do, and sign language, gesture, context, mutual karma, and "the language of love" will take one only so far, let's face it. One needs more than this to communicate as a full human being.

The New Language They Speak in Papua New Guinea

Thus in several locations in and around Oceania, people expanded and adapted this pidgin into a real language, called a creole language. Here is an announcement in creole version of South Seas jargon as used today in Papua New Guinea:

> Long dispela wik, moa long 40 meri bilong Milen Be provins i bung long wanpela woksop long Alotau bilong toktok long hevi bilong helt na envaironmen long ol liklik ailan na provins.

> Bung i bin stat long Mande na bai pinis long Fraide, Epril 22. Ol opisa bilong Melanesin Envaironmen Faundesen wantaim nesenel na provinsal helt opis i stap tu bilong givim toktok insait long dispela woksop.

What???! Here is the translation:

Long dispela wik, moa long 40 meri bilong Milen Be provins
in this week more than 40 woman of Milne Bay province

i bung long wanpela woksop long Alotau bilong toktok long hevi
he meet in a workshopin Alotau to talk about problems

bilong helt na envaironmen long ol liklik ailan na provins.
of health and environment in PLURAL small island and province

Bung i bin stat long Mande na bai pinis long Fraide, Epril 22.
meeting he was start on Monday and will finish on Friday April 22

Ol opisa bilong Melanesin Envaironmen Faundesen wantaim
PLURAL officers of Melanesian Environmental Foundation together with

nesenel na provinsal helt opis i stap tu bilong givim toktok insait
national and provincial health office he be also to give talk within

long dispela woksop.
in this workshop

"This week, more than forty women from Milne Bay Province are meeting in a workshop at Alotau to talk about the difficulties of health and environment in the small islands and provinces. The meeting began on Monday and will finish on Friday, April 22. The officers of the Melanesian Environment Foundation together with the national and provincial health office are there too to give talks in the workshop."

Unlike the pidgin passages from the 1800s, this passage expresses precise concepts with a systematic grammar. In other words, it is a language, called Tok Pisin (*Pisin* is from *pidgin*, by the way).

Inevitably, to us, Tok Pisin looks like baby talk–distorted-looking versions of English words strung together in ways that appear unfamiliar and thus "wrong"–apparently the product of just not knowing English and resorting to a ding-dong Rube Goldberg sort of substitution for grammar. Tok Pisin, however, is not "English"–it's a new language that happens to have recruited English words as its foundational material because of accidents of social history. Tok Pisin's *opis* for "office" is just one of thousands of sequences of sounds in the world's languages that happen to have come to be used to signify an office. *Opis* was *derived* from English *office*–but then to pronounce it "OH-pees" in conformity with the sound system of Melanesian languages is no more a "perversion" than the fact that English itself took *office* from French, where it was pronounced "oh-FEES."

Sometimes the English source of Tok Pisin words is obscured somewhat: *wanpela* for "a" comes from *one* plus *fellow*, *fellow* having been transformed into an ending signifying that a word is an adjective: note also *dispela* for "this." *Meri* for "woman" comes from *Mary*, a typical white woman's name (especially in the Victorian era when the English presence in Oceania was at its height–if the colonization began today, Tok Pisin for "woman" might be *Jennifer*!), generalized to refer to all women. *Ol* for "plural" is from *all*; *wantaim* for "together with" is from *one time*, as in "at one time." *Long* is from *along*, which was generalized into an all-purpose preposition like Russenorsk's *på* when the language was still a pidgin. Even a full language can get along very well with just one preposition (many do); in the creole, *long* does heavy duty, just as it did in the pidgin.

Where is the "grammar" in this language? For one, Tok Pisin has a regularized way of situating events in time. In one of the pidgin sentences, tense is left to context. The Melanesian says: *You plenty lie. You 'fraid me se-teal. Me no se-teal, me come worship. What for you look me se-teal?* where an English speaker would say *You're a big liar. You're afraid I'm* ***going to*** *steal. I don't steal–I* ***came*** *to worship. Why are you expecting me to steal?* In Tok Pisin, the *bye and bye* that one of the pidgin speakers used has been shortened and transformed into a rough equivalent of English *will*: *Bung . . . bai pinis long Fraide, Epril 22* "The meeting will finish on Friday, April 22." The status of *bai* as a piece of grammar is even clearer in this sentence from a father:

Pes	pikini	ia	**bai**	yu	go	long	wok . . .	**bai**	yu	stap	long	banis	kau
first	child	here	will	you	go	to	work	will	you	be	on	farm	cow

bilong	mi	na	**bai**	taim	mi	dai	**bai**	yu	lukautim.
of	me	and	will	when	I	die	will	you	take-care-of-it

> "You, first son, will go and work, you'll be on my cattle farm and when I die you'll look after it."

This man uses *bai* with every possible verb: in other words, just as we use *will*; *bye and bye* has been transformed from an adverbial expression into a regular marker of future tense.

Indeed, Tok Pisin has a whole array of ways to situate an event in time:

She goes to market.	*Em i go long maket.*
She goes to market (regularly).	*Em i **save** go long maket.*
She is going to market.	*Em i go long maket **i stap**.*
She has gone to market.	*Em i go long maket **pinis**.*
She went to market.	*Em i **bin** go long maket.*
She will go to market.	*Em **bai** go long maket.*

In the second example, *save*, which we saw in the pidgin used in the literal sense of *know* (*Tanna man no save work*) has developed into a marker of regular events (if you know an action, you probably do it or see it happen a lot). Earlier in the passage about the meeting in Milne Bay, it is mentioned that health and environment are central concerns for women in Papua New Guinea,

bikos	dispela	tupela	samting	i	**save**	kamap	strong
because	this	two	thing	he	(generally)	result	strong

long	sindaun	na	laip	bilong	famili	na	komyuniti . . .
in	situation	and	life	of	family	and	community

> "because these two things often have a strong effect on the situation and life of families and communities . . ."

That "false friend" *stap* in the third example now not only means "to be" as it did in the pidgin, but is used for the progressive just as English's *to be* is. *Pinis* is as fortuitously awkward in its shape as *Pisin*, coming from *finish* and here signifying that the event is completed, as, if you think about it, the *have*-perfect does: *Elvis Presley left the building* could refer to any time in Presley's life and could be said even if Presley had returned to the building since; *Elvis Presley has left the building* connotes that Presley just departed and is still departed. *Pinis*, then, is grammar–if you were to "make up" a silly, broken-down English, you can be pretty sure you would not include using *finish* as a way of distinguishing between *spoke* and *has spoken*. This feature in Tok Pisin is one of dozens showing that it is very much a systematic grammar, which must be learned and practiced like any other.

As a real language with a real grammar, Tok Pisin has also decided between being a "make it change" language versus a "changes itself" language, having settled on the former. Starting in the early twentieth century, a distinction arose between pairs like *boil* (what water bubbling on a stove does) and *boilim* (to make the water boil), *hariap* (to be in a hurry) and *hariapim* (to make someone hurry), and *lait* (what a light shining by itself does) and *laitim* (to make something shine – that is, to light it). The *-im* comes from *him,* but, obviously, English speakers do not use *him* in this way. Imagine mastering these pairs in trying to learn the language, especially as a native speaker, as you probably are, of either English, which doesn't regularly mark the "make it change"/"changes itself" distinction, or a "changes itself" language (for example, Romance, Germanic, Slavic).

Tok Pisin also has, as a "real language," arbitrary, culturally embedded conventionalizations of words. Is *perky* really the word we would apply to someone with, as a dictionary might say, "a buoyant or self-confident air; briskly cheerful"? Michael Richards' *Seinfeld* character Cosmo Kramer was buoyant, self-confident, and briskly cheerful, but one would not call him "perky." On the contrary, *perky* in American English is a word that one cannot use without quotation marks–it has taken on a deeply ironic tone, is largely restricted in regard to gender (referring usually to women), and carries a faint whiff of disapproval in implying a certain shallowness. Many words in a real language are like this; you kind of have to "be there" to fully grasp their arbitrarily particular meanings.

Words in Tok Pisin have developed shadings like this–bread-and-butter stuff like *man, meri, stap,* and *pinis* are only the outer layer of the language. In one area, for example, *orait* meant not only "allright" as in "fine" but also "equal to the white man in regard to knowledge," obviously a highly culturally embedded connotation for an originally rather mundane word. *Hevi* from *heavy* started in its literal meaning but now extends into the realm of sadness: in the passage on pages 140–141, *hevi* is the established word for *difficulties.* For example, *bel bilong mi i hevi* "belly of me he heavy" means not that one is full of food or has unusually weighty intestines but that one is sad.

The status of Tok Pisin as a true "language" becomes especially clear in how it is spoken by children for whom it is not a second or third language but their primary, native one. This is increasingly true for children born to parents speaking different languages who have moved to a city where, because no one of dozens of languages has any reason to be adopted by the general population, Tok Pisin has become the public vehicle of general communication. Urban children in Papua New Guinea may learn one or both of their parents' languages but just as often speak them only partially if at all and often grow up most comfortable in Tok Pisin. Urban, natively spoken Tok Pisin differs from rural, second-language Tok Pisin in various ways. One of them is that it is spoken faster, which encourages the clipping and squashing of words. The second-language speaker will say *Mi go long haus* for "I am going into the house," whereas the urban child will say *Mi go l'aus*; the adult will say *wanpela man* for "a man," the child is more likely to say *wanla man*, and so on.[3] In other words, among urban native speakers of Tok Pisin, the language is doing something inherent to all full languages: it is *changing,* just as *æfre ælc* became *every* and *de de intus* became French's *dans*. Having started down the road toward eventually evolving into a new language is yet another way in which Tok Pisin has joined the "language" club.

In regard to taxonomy, creoles are definitely distinct "languages" from the ones from which they take their words. We have seen that nonstandard dialects of a noncreole language can diverge

---

3. Some readers may be familiar with the oft-encountered claim that a pidgin becomes a creole only when children acquire it and may wonder why I have described adults as transforming the language into a creole. The adult: pidgin/children: creole formula is in fact an oversimplification: although there have been a very few cases in which children have been responsible for turning a pidgin into a full language (as will come up in a different context in Chapter 5), in most recorded cases, adults accomplished the transition from pidgin into creole long before significant numbers of children even acquired it. Acquisition by children certainly lends a "finishing gloss" to all of the features distinguishing a lingo from a language (as we see with Tok Pisin), but the difference between such languages as used by adults as second languages from the varieties used by children as first languages are hardly stark enough to serve as a useful taxonomic yardstick between pidgin and creole. For example, long before urbanization led large numbers of children to use Tok Pisin as a primary language, it had already developed all of the grammatical features I have described and a great many more.

quite sharply from the standard ones we see and hear most often. Creoles, however, are a case apart from even the most divergent dialects of their "host" languages. In Standard French, *Where were you?* is *Où étais-tu?* In the French spoken by Québecois Canadians, however, this would be *Où c'est que t'étais, twe?* This is different from the Parisian sentence but recognizable as "French"–all of the words are used in ways processible by a Parisian speaker; it's just that Parisians do not happen to use them in all of those particular ways, just as if we hear an upcountry British waitress say, *What is it you'll be havin', then?* we are faced with a quick adjustment, not bafflement. In the creole French of Martinique, though, the sentence is *Weti ou te ye? Weti* is simply a word for "where," but originated from the French sentence *Où est-il?* "Where is he?"; *ou* is shortened from *vous* (pronounced "VOO"); *te* is from *était* "was"; *ye* is from *est* "is." Clearly, we depart here from what anyone would classify as "French"; this is the raw stuff of French retooled to new ends.

Creoles: Louisiana Is Just One of Many

Tok Pisin is one of several dozen creole languages,[4] all of which developed when people were led to expand a pidgin into a full language. This is a rather unusual circumstance and was conditioned most often when people were marooned for several years or permanently in a context where several languages were spoken, such that the only practical solution to communication was to use an incompletely acquired version of the language spoken by the dominant group. As a result, most creoles formed during the so-called "exploration" of the world by a few European powers from the 1400s through the 1800s, in which cultivation of food and material goods to enrich the coffers of the "exploring" country required large crews of manual laborers to do work that whites back home were only fitfully willing to do. Namely, the slave trade and its con-

4. How many creoles one considers there to be depends on where one draws the line between "pidgin" and "creole" as well as between language and dialect. If we treat *creole* as referring to languages with a systematic grammar that serve the needs of full communication (although not necessarily being first languages for all or most of their speakers) and take a "lumper" rather than "splitter" perspective in distinguishing dialects from languages, by my count there are about sixty creoles worldwide.

tractual aftermath under a different name in the 1800s gave birth to several dozen creoles.

Surinam, on the northern coast of South America south of the Caribbean, was developed as a slave colony by the British in 1651, and Africans there developed a creole called Sranan. Today spoken not only by blacks but by whites and even the descendants of Indians and Chinese encouraged to immigrate as laborers in the 1800s after the demise of the slave trade, Sranan, like Tok Pisin, has harnessed English words into a structure of its own. *A hondiman dati ben bai wan oso gi en mati* is, believe it or not, derived entirely from English words:

A	hondiman	dati	ben	bai	wan	oso	gi	en	mati.
the	hunter-man	that	PAST	buy	a	house	give	his	mate

"That hunter bought a house for his friend."

The home languages of the British West Indies also are creoles, such as Jamaican patois. There also is a related creole on Barbados, and because the first slaves brought to Charleston, South Carolina, in the late 1600s were brought mainly from Barbados, today blacks on the Sea Islands off the coast of South Carolina speak a creole called *Gullah,* which is an offshoot of West Indian *patois* spoken here in America.

Not only English has been pidginized and creolized; there are creoles using words from many languages. One of the obscurest places on earth, almost never featured in the press, whose inhabitants few of us ever meet, are the islands off the west coast of Africa in the Gulf of Guinea: São Tomé, Principe, and Annobón. These islands were among the formerly unknown territories encountered by Portuguese explorers on their sail down the African coast starting in the 1400s, and they soon established sugar plantations on the islands, worked by Africans imported from the nearby coasts. An almanac will tell you that the official language on these islands is Portuguese, but the native language of the ninety percent of the population descended from these slaves comprises various dialects of a Portuguese creole, which has molded Portuguese words into a new framework so thoroughly that, if one were not told that the language had a historical relation with Portuguese, one might listen to it for weeks without even noticing the relationship even if one were a native Portuguese speaker. Even in a simple sentence like this from the Annobón dialect, Romance language speakers instantly know they are not in Kansas anymore. In Portuguese *The children ran there* would be *As crianças correram para aí*; in the creole,

Namina nensai xa sa xole ba iai.
child those run go there

Namina is derived from *menino*, a word for "child" now used in Brazil, having been replaced by *criança* in Portugal itself. *Nensai* is a combination of an African word meaning "those" and Portuguese's *essa* "that." I did not give translations for *xa* and *sa*, because their meaning is too subtle for any one-word translation to be appropriate. Suffice it to say that *xa* is derived from *cá* "here" and *sa* from *são*, one form of "to be," and that together they serve to show that the event is being seen from the outside and had a definite end rather than continuing indefinitely (such as, say, *The children were running there*). That there exists a regular way of expressing that nuance is one more indication that creoles are real "languages." The first sound in *xole* is pronounced rather like the *ch* in *Bach*, and the word is derived from *correr*: it's not a long way from the *k* sound of the initial *c* in *correr* to the *ch* in *Bach*; the first *r* became an *l*; the final *r* was eliminated. *Ba* is from one Portuguese form of "go," *vai*. *Iai* is from *aí* "there." Obviously, however, the roots of this sentence in *menino, essa, cá, são, correr, vai,* and *aí* are merely academic today; these words have served as the rootstuff of the birth of a new language that follows a plan quite unlike Portuguese or any other Romance (or European) language.

Before they got down to São Tomé, the Portuguese had also settled the Cape Verdean islands, and here, too, a Portuguese creole is spoken. By the early 1500s, the Portuguese had reached India and left creoles along the western coast of India and in Sri Lanka; people of Indian parentage occasionally mention that their grandparents speak "Portuguese"–as often as not this is one of these Portuguese creoles. The Portuguese also went on to leave creoles in parts of today's Indonesia and as far as Macao.

In the popular consciousness, "creole" is often associated with *café-au-lait*–colored people in Louisiana and the Euro-African hybrid culture connected to it, and as such it is commonly supposed that *creole* refers to a language spoken in Louisiana. As we have seen, a creole is a *type* of language, not the *name* of just one. But there is indeed a creole language spoken in Louisiana, by descendants of African slaves brought there when it was a French territory. Here, French has undergone the same process of profound reinterpretation as English and Portuguese did in becoming creoles. Here is a passage from a description of how Africans made drums:

Ye	te	konnen	pran	en	bari,	avek	en	but	lapo
they	PAST	know	take	a	barrel	with	a	piece	skin

e	ye	te	gen	shofe	lapo	pu	li	vini	stiff.
and	they	PAST	have	heat	skin	for	it	come	stiff

"They used to take a barrel with a piece of skin, and they used to heat the skin until it became stiff."

Haitian Creole is another French creole, having developed when Haiti was taken from the Spanish by the French and devoted to sugar cultivation. Related creoles are spoken on other French Caribbean islands such as Martinique and Guadeloupe, as well as down in French Guiana. Meanwhile, the French also established plantation colonies on the other side of the world on the islands of Mauritius, Réunion, and other smaller ones nearby.

Papiamentu of Aruba, Bonaire, and Curaçao is a Spanish creole, and there is also a Spanish creole spoken in an isolated village in Colombia where escaped slaves established their own community, which lives to this day, and a Spanish creole dialect cluster spoken in the Philippines. The Dutch concentrated more on supplying slaves to other powers than on running plantation colonies themselves, but slaves did creolize Dutch a few times, on the Caribbean island of St. Thomas and in two locations in Guyana (each of these creoles is now extinct or all but).

Because colonization was such a fertile ground for creole languages to emerge, there are no creoles based on the languages of the European powers that did not extend their domains overseas and recruit multiethnic work crews to cultivate them. Russia extended eastward across Siberia rather than overseas and did not herd the peoples they encountered into multiethnic work camps; thus there are no Russian creoles (although there have been some Russian pidgins, including one between Russian and Chinese!).

Similarly, there is no Italian creole.[5] By the time Germany, only unified in 1871, attempted to gain a toehold in overseas possessions, where they gained dominion, pidgins or creoles based on other languages already existed (as in Samoa) or some other interethnic lingua franca was already long established (such as Swahili in East Africa). The lone German creole arose in the unusual circumstance of a single orphanage run by Germans in Papua New Guinea, where children of a casserole of ethnicities (Papuan, Chinese, Malay, English-Samoan halfbreed, English, Australian, and German) developed an in-group creole German called Unserdeutsch ("our German").

Yet some creoles based on non-European languages have arisen. In the late 1800s, men speaking a number of African languages were recruited into an Egyptian army occupying what is now southern Sudan. The men were commanded in Arabic, which as a pidgin became the vehicle of communication for the army just as South Seas Pidgin English had been on plantations in Oceania. Expelled by a Mahdist insurgency, the Egyptian army was given haven by the British in Uganda and Kenya. The Arabic pidgin, used every day from the start and then rendered a lifelong primary language for the soldiers, expanded into an Arabic creole still spoken by the soldiers' descendants.

In the Pacific Northwest and British Columbia of the late 1700s, the lingua franca between whites and Native Americans, as well as between Native Americans of different ethnicities, was a pidgin version of the Chinook language. Starting in the mid-1800s, according to the imperatives of Manifest Destiny, Native Americans in Oregon were relegated to reservations, where people with various native languages spent the rest of their lives in cohabitation. This, like the Sudanese army context, was analogous to a plantation in regard to language use, and on one of these reservations, called Grand Ronde, the Chinook pidgin (called Chinook Jargon) was developed into a short-lived creole.

---

5. In the strict sense, this is a good thing, because the tragic truth is that most creoles have arisen amid conditions of unthinkably stark and ineradicable social injustice. However, if we can allow ourselves to very briefly take a purely hypothetical perspective, a creole based on an encounter between Italian and West African languages would most likely be a ravishingly beautiful tongue to both Western and African ears.

Slippage Between Folk and Scientific Terminology

Sometimes, the status of a language as a creole–a true language–rather than a pidgin–a broken version of another language–is obscured by the typical lack of fit between folk and linguists' usage of the word *pidgin*.

Hawaii, for instance, is home to an English-derived creole. In the late 1800s, American business interests could not bring Africans to these islands to grow fruit for Dole and other companies, and so they first imported Chinese laborers. After America's Chinese Exclusion Act of 1882 virtually cut off all Chinese immigration, waves of Japanese were recruited in their place. After the Japanese were similarly banned by the Gentleman's Agreement of 1908, Filipinos and some Koreans and Puerto Ricans were recruited, all mixed with native Hawaiians and the Portuguese imported as plantation foremen. Most of these people were technically signed on temporarily, but quite a few never earned enough money to go home or they found conditions in Hawaii more congenial than limited opportunities in their homelands. The pidginized English that this mixture of people naturally developed expanded into a creole at the beginning of the twentieth century, but tradition has maintained the language's local name as "Pidgin."

Another example is West African "Pidgin" English. One might wonder what happened to the slaves in America who fought on the British side, as many did. As it happened, England relocated many to Sierra Leone (by the initial destination of Nova Scotia, whose climate many slaves found–big surprise–a little discomfittingly chilly). Shortly afterward, Jamaicans descended from escaped slaves were brought to Sierra Leone as well to live free. This mixture of immigrants, some speaking a form of Gullah and others Jamaican patois, gradually created a new offshoot of West Indian patois called Krio. Krio-speaking administrators were placed in other colonies down the West African coast throughout the next century, such that today variants of the same creole are spoken in Ghana, Nigeria, and Cameroon. Yet in these three countries the language is called "Pidgin," which implies that it is merely "bad English" along the lines of the Melanesian citations from the 1800s, this being reinforced by the fact that Krio and its descendants are rarely written.

From Pidgin to Creole: Just Add What?

When people "expand" a pidgin into a creole, what kind of "material" do they use? How do they "know" what structures to add, what words to refit to express aspects of the grammar necessary to a full human language?

A Different Kind of Mixture

A major source for creole structures is, as one might expect, the languages that the people creating creole speak natively. Much of what distinguishes a creole as being built on a "different plan" from the language that provided its words consists of sound and grammatical patterns from these natively spoken languages.

Tok Pisin and other creoles based on South Seas Pidgin English attach *-im* to a verb when it is followed by an object (*boilim wara* "to boil water") because Melanesian languages have this same pattern. Here is a word-by-word comparison of *I didn't see your pig* in another creolized offshoot of the old South Seas Pidgin, the misleadingly named Pijin of the Solomon Islands, and one language that its creators spoke, Kwaio:

Pijin:	Mi	no	luk-im	pikipiki	bulong	iu.
Kwaio:	Ku	'ame	agasi-a	boo	a-	mu.
	I	not	see-it	pig	of	you

Because of parallels like these, Solomon Islanders find Pijin easy to learn, more so, for example, than we would. On the subject of *we*, *we* seems to be a pretty simple word to an English speaker, but the South Seas English creoles distinguish between two kinds of "we-ness" in ways that would never occur to most of us. *We* in the Solomon Islands variety is *iumi*–"you-me"–but only if it refers to the speaker and the people being spoken to; if the *we* refers to the speaker and some people not in the conversation, then it is *mifala*–rather like "me and those fellows." But if the speaker is referring to himself and just one person not in the conversation, then *we* is *mitufala*–more or less "me and that fellow, us two." Finally if the speaker is referring to himself and a *single* person he is speaking to, then *we* is *iumitufala*–"you and me two fellows." Again, this pattern follows directly from the way pronouns work in many of the indigenous languages of Oceania.[6]

When the Surinamese says in Sranan *A hondiman dati ben bai wan oso gi en mati*, he says that the hunter "bought a house gave his friend" instead of "bought a house for his friend"; the person speaking the Portuguese creole of Annobón says that the children "ran went" to a spot. These people are using a pattern inherited from West African languages that many of their creators spoke, which rely less on prepositions than European languages do and instead tend to run verbs together to express what the typical European language uses words such as *to* and *for* for. In Fongbe, spoken in Togo and Benin, to say *Koku brought the crab to her*, you say:

Koku so ason o na e.
Koku take crab the give her

Chinese speakers will find this way of arranging words familiar; it's one of those grammatical features that some languages have and some don't. The region of Africa where languages have this feature is actually rather small, but because this region happens to be coastal and easily accessible by sea (the eastern side of Côte d'Ivoire through Ghana, Togo, Benin, and Nigeria), the European powers drew many slaves from the area, thus ushering this verbal quirk of one subgroup of African languages into all of the Caribbean creoles on the other side of the world.

Creole sound systems also echo the ones in their creators' native languages. Most West African languages tend to follow any consonant with a vowel, even at the end of a word, such that their words are all couched in a consonant-vowel-consonant-vowel pattern. You can see this tendency in the Fongbe sentence. This is found in many languages, Japanese being one, which is why in Japanese *ice cream* comes out *aisu kuriimu*, with vowels plugged onto the ends of both words and even between the first two consonants in *cream*, to make the word conform to the Japanese sound pattern.

---

6. I can think of no region on earth where languages have more beautiful names than the South Pacific. Some others: Roviana, Erromangan, Nguna, Tangoan, Tokelauan, Rapanui, Kiribati, Malekula, and my favorite, to me less pretty than just neat, To'aba'ita, where the apostrophes stand for glottal stops, such as the "catch in the throat" before the two syllables of *Uh-oh*.

This same feature in African languages is why *that* in Sranan is *dati*, *mate* is *mati*, etc., whereas in the Portuguese creole of Annobón *correr* becomes *xole.*

Intertwined Language versus Creole

This kind of mixture may justifiably lead the reader to ask what the difference then is between creoles and intertwined languages such as Media Lengua and Michif. Creoles, unlike the intertwined languages, are not straight cocktail-style blendings of one language with another. On the contrary, the language that receives "mix-ins" from other languages is a *pidgin* version of the language being learned and, because the pidgin is a simplified system, the people transforming it into a creole tend to mix even the features of their native languages into it in *pidgin* form. For example, Oceanic languages tend to distinguish not only "we two" from "we all," but also "we three." No dialect of the South Seas English creoles goes this far, however–Melanesians in effect contributed a pidginized version of their own language into the new language.

Another thing that put a block on how much speakers could transfer from their own language into the creole is that there were usually several other languages spoken besides their own. All of the men who created Media Lengua spoke one language natively, Quechua. This meant that they could attach all of Quechua's specific endings onto Spanish words and use the words in a grammar paralleling Quechua's in great detail, because such a mixture would be comprehensible to all of them. On a Caribbean plantation, however, the slaves, drawn from a stretch of African coast that today extends from Senegal down to Angola, often spoke as many as a dozen different languages. In each language, a given aspect of grammar was often expressed in different ways. Take the past tense: the Twi language uses a prefix; Mandinka, Igbo, and Kikongo use a suffix; Wolof and Yoruba have a separate particle that comes before the verb; Ewe uses the verb *finish* placed after the verb. Moreover, even when, say, both languages use a suffix, they use different ones: the past-marking suffix in Mandinka is *-ta*, but Igbo's is *-le*, and so on. A direct mixture of so many languages, with speakers of each randomly tossing their own prefixes and suffixes and structures into the pot, would be a ragtag mélange useful to no one.

Imagine if English speakers were marooned on a plantation run by Spaniards, with just as many Japanese, Zulu, and Arabic speakers. If after everyone had learned a few hundred words of Spanish, English speakers attempted to run around saying things like "The mujer habla-ed Español pero the hombre is hablaing Inglés," English speakers wouldn't have any trouble with it but the Arabs, Zulus, and Japanese would only get the Spanish parts; only the Arabs would be able to even venture what *the* was for, because Japanese and Zulu have no equivalent words.

Thus speakers tended to bring features into the creole that were common to most of the languages. The running of the verbs together, for example, is found in Twi, Ewe, Yoruba, and Igbo and in many other West African languages and thus would have been familiar to a majority of speakers on many plantations (and would not have been too difficult to handle for people speaking languages that did not have this feature). Thus a Caribbean creole is not, properly speaking, "an African language with European words," because a great deal of the grammar of any particular African language did not make it into any creole.

Footprints from the Past and Tokens from Beyond: The Dominant Language's Imprint

Because the creators of a creole cannot transfer their entire native grammar into the language, there is plenty of room left over for structures, rather than just words, from the language being learned as well. Yet the languages that creole creators were exposed to were not the versions of those languages most familiar to us today.

For one, because most creoles formed a century or more ago, this means that the version of a language from which a creole took its words was an earlier stage of that language: Sranan, for example, developed only thirty-five years after Shakespeare's death. Furthermore, in the Caribbean cases, the whites with whom Africans had the most contact were not the ones speaking the standard dialect of the language. In the 1600s, today's situation, where almost every inhabitant of a given European country can at least handle the standard dialect, had yet to arise. At the time, most people who used the standard dialect as a home variety lived fairly comfortably in the geopolitical heart of the country and had little reason to leave their stable lives in a European city to carve out a life on a blazing hot, subequatorial island or coastline. Whites who ran plantations or contracted themselves out to work them were generally people of what used to be called "the lower orders" and, as such, were speakers of various regional or nonstandard dialects or both. We see a latter-day extension of this in depictions of the *Titanic* tragedy, where the third class included a large number of Irish people seeking work in America, especially as servants.[7] Equally large numbers of Irish and Scots-Irish, speaking distinctly nonstandard varieties of English if they spoke much English at all, had immigrated to the United States and other colonies as early as the 1600s. The dialects that these people spoke had a major effect on how English creoles came out, as did nonstandard dialects spoken by French, Portuguese, Dutch, and Spanish colonists.

In the French creole of Mauritius, *They were going* is *Zot ti pe ale.* This looks nothing like what we know as "French," where the sentence would be *Ils allaient*, and tempts us to suppose that the sentence must be modeled on the African languages that Mauritian slaves spoke. Yet the sentence is much more familiar from the perspective of French dialects spoken beyond Paris. In many regional French varieties, *they* is not *ils* but *eux-autres*, literally "them others" (Spanish's *nosotros* began similarly as *nos otros* "we others"). In rapid, casual speech, the full pronunciation of *eux-autres* ("oo-ZOAT-ruh," the *oo* being the one in *foot*) is elided to simply "ZOAT." African slaves in Mauritius mostly heard nonstandard dialects of French, and thus it was natural that they brought the French pronoun into the creole as *zot*. Another construction in many regional French dialects is the use of *après* "after" in a progressive construction: in Montreal today one can easily hear a French Canadian say something like *Le chat était après jouer* "The cat was after playing" to mean *The cat was playing. Après* is pronounced "ah-PRAY"[8] and, once again, it was quite natural for Africans hearing this used in rapid speech to process it as something like "PRAY" and subsequently,

---

7. Did you know that passports were not required for travel to foreign countries until after World War I? That's why, as late as Henry James, characters never mention them and why they go unmentioned in accounts and depictions of the *Titanic* disaster.

8. Think of the space I could save in giving these pronunciations if French's spelling system weren't such a mess! Of course, an English speaker saying this is a prime example of a pot calling the kettle black.

through their own casual speech, transform it further to *pe*, just as we tend to say *Djoo see?* instead of *Did you see?*

Now we are in a position to see the source of *Zot ti pe ale*—in essence, it is merely a perfectly ordinary sentence of non-Standard French, rendered to manifest the realities of running speech and remodeled slightly by the fact that the Africans were exposed to French as adults learning a second language. *Eux-autres étaient après aller* in rapid French comes out roughly as "zoat t'pray ah-LAY"—and there is only a short jump from here to *Zot ti pe ale*. This is, of course, a "cooked" example; there is a great deal in Mauritian creole French that does not trace back to any kind of French this directly. Yet such cases are numerous enough to constitute a central component of creole grammars.

In Gullah of South Carolina, there is the word *blant*, which is yet another form showing that an action is habitual: *Uh blant come yuh ebry eebnin* "I come here every evening." Outwardly a random oddity, *blant* actually has its roots in the *belong* construction of dialects of southern England such as the Cornwall one we saw in Chapter 2: *Billee d' b'long gwine long weth 'e's sister* "Billy goes with his sister." People from this region were numerous among those who left England to work on plantations, often alongside blacks, in America as well as Caribbean colonies. Because of features like these, a creole often serves as an interesting window on its "host" language's past as well as its range of dialectal variety.

The Ancient Verities of Human Language

Finally, creole creators round out their new language by utilizing what can be considered "default" settings available to a human grammar.

All of us have a callus on the inner side of one of our middle fingers, from a lifetime of holding writing implements.[9] Obviously, this callus is not genetically specified; we are not born with it, and people who have never written do not have this callus. Yet the callus does no harm; it is a conditioned add-on to the constitutions

---

9. I originally had this as "on the left side of our middle finger" but was correctly informed that this was an example of galloping righthandedocentrism (thank you, Erika).

that we simply go through life with.[10] My cat, with her needle-sharp claws, astounding physical agility, and instinctive vigilant crouch whenever she spots a bird, is clearly genetically programmed to espy, pounce on, and kill prey. Yet she has spent her life living with me, eating food from a bowl, having never killed anything (she has yet to bring me a dead bird the way one hears of house cats doing), and in general leading a quiet, indolent life focused on intently watching everything I do and getting me to pat her as often as possible. She is quite functional and happy, even though this life style in the technical sense requires a lifelong suppression of her native predilections.

In the same way, many linguists have argued that there are "default" linguistic structures, which most individual languages retain only a subset of, language change having altered and hidden the others through the millennia. Under this analysis, human cognition easily copes with these deviances from underlying "preferences." One of the proposed universal settings is the word order subject-verb-object, which has been argued to be the most "natural" order, given that the description of an event requires first situating it in regard to the actor, then designating the action itself, and then designating what entity the action affected. To an English speaker nothing could seem more "normal," but then to a Hindi speaker, whose verbs come at the end of the sentence, or an Irish speaker, whose verbs come at the beginning, this conception might seem somewhat "linguocentric." There are even a few languages where the object comes first, then the verb, and then the subject!

But creoles are one piece of evidence that subject-verb-object order is indeed a "default," in that creoles tend to have this order regardless of the word order of the languages spoken by their creators or the language that they borrowed the creole's words from. There is a virtually extinct Dutch creole spoken in an isolated community in Guyana, only discovered in the late 1970s when one creolist was told that there were some people up the river who spoke a "funny kind of Dutch." Dutch, a close relative of German, places its verbs at the end of sentences in many cases. Most of the creators

10. Note also that, for many of us, this callus has gotten thinner and less distinct in the past decade as we do more and more of our writing on computers.

of the creole spoke an African language called Ijo (actually the *Eastern* Ijo dialect), which also places its verbs at the end of sentences. Yet the creole, Berbice Dutch, has subject-verb-object order:

En mo-te an invai-te eni frendi-ap.
they go-PAST and invite-PAST their friend-s

"They went and invited their friends."

Creoles also suggest a "default" setting in how they indicate tense. The African languages that slaves brought to the Caribbean had various ways of expressing past tense: some with a prefix, some with a suffix, some with a particle before the verb, some with an element after the verb. In Sranan, past is expressed with a particle placed before the verb–*A hondiman dati* ***ben*** *bai wan oso* "That hunter bought a house"–as is the progressive and the future and most related concepts. In this respect Sranan parallels most creoles, where regardless of how the dominant group's languages or the native ones expressed tense, the creole does it with particles placed before the verb. Noncreole languages have often strayed from this setting over time: English, for example, has *will* as a preverbal "particle" to express the future but has a suffix for the past (*-ed*) and the progressive (*-ing*).

Creoles, then, are mixed languages of a sort. However, more specifically, they combine pidginized elements from the creators' and the dominant group's languages and then expand this into a true language. The expansion is accomplished largely through refashionings of features already in the pidgin, this directed in part by universal "default" constructions that older human languages have often drifted away from.

Is It a Creole or Not? A Matter of Degree

Like language change, the language–dialect distinction, and degrees of language mixture, the extent to which a language is pidginized before it becomes a full language again is a matter of degree rather than stark metrical distinctions. In the most "creole" of creoles such as Tok Pisin, Sranan, and Annobonese, a language was reduced to its barest of bones, little more than a collection of words and their basic order, and built back up into a new language.

In other cases, social conditions will allow a group of learners to acquire somewhat more than a pidgin version of a language but still not attain full competence. The language will thus be reduced to a considerable, but lesser, extent, retaining perhaps a few of its endings, a smattering of its irregular forms, and some of its more unusual sounds. The people who first create this version of the language can be said to "speak" it–just not fully, rather the way many adult Latino immigrants living in predominantly Spanish speaking communities speak English.

The Languages Formerly Known as Creoloids

Yet if social conditions lead speakers to use a language in this form every day and these conditions are such that this form of the language is what is passed on to new generations, then this "almost there" variety acquires its own rules, filling in its structural gaps, and it becomes a language in its own right. Because the creators of these languages acquired from the start a less pidginized version of the dominant language than did Melanesians in the South Pacific or Africans in Surinam, there is less room for carryover structures from the creators' native languages. The result is a language straddling the line between "dialect of X" and "creole with words from X"–a *semicreole.*[11]

A classic case here is the French of the small island of Réunion, east of Madagascar. In the late 1600s, French colonists established small farms here and imported slaves from Madagascar. The slaves often lived in the white master's house, sometimes even all sleeping in the same bed as the master. Close conditions like these allowed the Malagasies to acquire more than a pidgin French (as many of us know, a foreign language is learned best on the pillow!), but then acquiring full French was still a distant prospect. First of all, people's ability to completely acquire a language atrophies significantly after the early teens, as we see among, for example, the increasing numbers of Russian immigrants today. Russians who came to the

11. In the 1970s, when these languages were identified as a class, it was proposed that they be called *creoloids*. This never caught on, I suspect in part because about this time *Star Wars* came out and institutionalized words like *android*, which lent the *-oid* suffix an air of gimmickiness and triviality that rendered *creoloid* rather ineffably dopey.

United States as small children speak perfect English (and often make small mistakes in their Russian), ones who came in their twenties or later speak with a definite accent and make occasional mistakes, whereas those who came somewhere in their mid-teens often have a slight accent and make the very occasional little slip-up especially when they are tired. Most of the Malagasies were adults and were thus hindered from acquiring perfect French first by sheer age.

Then there was the sociological factor: if one, say, marries a French person and relocates to France, one generally makes one's best efforts to speak French as well as possible even as an adult. Imagine, however, having been taken hostage by the Taliban in Afghanistan and kept in a small village. One's desire to acquire the local language would have been diluted by a sense of one's captors as a nemesis; one may well have gone through the rest of one's life speaking functional but choppy Pashto or Arabic regardless of one's linguistic talents. Along these lines, the Malagasies acquired something recognizable as "French," but French only partially acquired–a semicreole French.

However, if the second generation of Malagasies (or, inevitably, Franco-Malagasies) had grown up in the same setting, then they would have had enough contact with whites to acquire it (that is, the local nonstandard variety) perfectly, just as immigrants' children in the United States grow up to speak perfect American English rather than inheriting their parents' accents and mistakes. But in 1715, the island switched to coffee cultivation, which required an increase in manpower. The French turned to East Africa for the vastly larger numbers of slaves needed, and soon Réunion became a typical plantation colony, with small numbers of whites dwarfed by large numbers of brown-skinned people living separately from them in a context of overt interracial hostility and mistrust.

Thus the first generation of blacks born on Réunion were waking up not in the whites' homes but in slave quarters. Yet even here, unlike South Seas indigenes or Africans in Surinam, they and their parents had at their disposal not just brief, utilitarian conversations with whites, but interaction with Malagasies who had acquired a semicreole French. Hence the semicreole became the local language. Here is a sample:

Alor mon papa la tuzur di amwen, en zur kan li lete
then my father PAST always say to-me one day when he was

zenzan, alor parmi de kamarad i di kom sa . . .
bachelor well among some friend he say like that

"Well, my father always said to me, one day when he was a bachelor, well among some friends he said like this . . ."

(I'll leave you in suspense as to what the father said!) There are definitely creole-style features here. *Li* for *he* on the first line is from *lui* instead of the subject form *il.* There are no endings for marking tense; instead, tense is indicated with particles before the verb. Although we do not see it here, verbs are in the same form in all persons and numbers.

Yet the roots of this variety in French grammar are clear. There is nothing here as foreign to the Western European eye as Tok Pisin's *givim toktok* for "give a talk," *dispela* for "this," *save* as a marker of habituality. All of the words are used in ways recognizably derived from their usages in French; even the initially puzzling *zenzan* is from *jeunes gens* [zhuhn ZHAW^{ng}] "young people." There are even some dialects with an ending for past tense and one for future: *mâz* "eat," *mâze* "ate," *mâzra* "will eat." A person who speaks French, taking into account that Parisian French would look more like Réunion French on the page if its spelling were more phonetic, could easily piece out the passage just cited. On Réunion, pidginization only ever went so far, and thus the creole is not as far from French as Tok Pisin is from English. Is this a dialect of French or a creole? There is really no answer–Réunion French is a blue-green crayon in the Crayola box.

Because the social conditions leading people to refashion an incompletely acquired language as a natively learned one have arisen most often in contexts of forced migration, most semicreoles are, like most creoles, spoken by brown-skinned people. Yet the birth of a creolized language stems in essence from the ineluctability of humans' innate capacity for full language, and as such there is nothing inherently "Third World" about the process. Afrikaans of South Africa, for instance, is a semicreole.

At first glance, this language, spoken by South African whites (as well as Coloureds and many blacks) seems to be simply a variant of Dutch, and indeed, when it arose after the Dutch colonization of the Cape of Good Hope in 1652, it was called "Cape Dutch." Yet Dutch did not simply evolve uninterrupted in the new colony the way English did in the United States or French did in Canada. Click-language-speaking Khoi tribesmen, or "Hottentots," were hired as nurses and herdsmen, and Dutch colonist men often had children with Khoi women. The Khoi learned a pidginized Dutch, and not only were their mixed children often exposed primarily to this rather than European Dutch, but even white children, often essentially raised by Khoi nurses, took this pidgin Dutch as their primary language (just as in Gullah-speaking country in America, white children raised by black "mammies" often used to speak Gullah as well as English). Quite soon, Dutch in the colony overall had acquired a distinct character of its own. Because the colonists themselves spoke Dutch and Standard Dutch was the language of education, white Afrikaner children did not fashion their new language *only* from pidgin Dutch. However, the "pull" of the pidgin was enough to deflect them from engaging only the Dutch of Holland; instead, the combination of Holland's Dutch, the Khois' pidgin Dutch, and features from Khoi itself created a Dutch semicreole.

During the reign of apartheid, this aspect of the roots of Afrikaans was resisted by South African philologists, uncomfortable with the idea that contact with blacks played a pivotal role in the formation of their language. But the evidence is crystal clear and no longer disputed today by any serious pidgin/creole expert. Dutch nouns have two genders; Afrikaans nouns have just one. Dutch verbs take different endings according to person and number; Afrikaans verbs, like creole verbs, remain unchanged regardless of person or number. Dutch, like English, has a past tense formed either by adding a suffix or changing a vowel and perhaps more: *ik zei* "I say," *ik zeide* "I said," *ik zie* "I see," *ik zag* "I saw." All of this goes out the window in Afrikaans, which instead uses only a construction with *have* as in *Ek* ***het*** *die glas* ***ge****breek* "I broke the glass," and one can count the irregular verbs on one's fingers. We English speakers should be so lucky as to find a European language presenting so little of the bric-à-brac and noisome exceptions that bedevil us left and right in tackling French, Russian, and, indicatively, Dutch itself.

To be sure, Afrikaans is a semicreole rather than a creole. We see one indication of this in the sentence just given, where the verb comes at the end as it would in Dutch, rather than the "default" subject-verb-object order that a "deep" creole, such as Berbice Dutch, has. Yet the weight of general contrasts with Dutch, and specifically the simplificatory leaning of those contrasts, can trace only to a degree of pidginization in Afrikaans' history. This is made even clearer in some Khoi features in Afrikaans, such as double negation in sentences like *Sy eet* ***nie*** *pap* ***nie*** "She doesn't eat porridge," paralleled by an identical negation construction in Khoi.

Every String in the Bow: Creole Continua

In some creolophone societies, an entire continuum of varieties is spoken, ranging from a "deep" creole through semicreole varieties into the local standard dominant language–all coexisting in one country. In Guyana, for example, as Derek Bickerton in his early work noted, one can ask a class how to say *I gave him* and receive fourteen different answers, representing all of the levels of a continuum of pidginization:

mi bin gii am	a di gii ii	a giv im
mi bin gii ii	a did gi ii	a giv him
mi bin gi i	a did giv ii	a geev ii
mi di gii ii	a did giv hii	a geev him
mi di gi hii	a giv ii	*I gave him*

A creole continuum like this resulted from the social stratification of slaves in plantation societies. Slaves who worked in the fields had the least contact with English and thus worked from the most reduced pidgin English to build a full language, the result being a "creole" creole. Other slaves, however, worked as artisans of various kinds or as overseers and thus had more contact with whites, enough to develop a semicreole variety. Slaves who worked in the house or in houses in the cities had even more contact with English and thus developed a semicreole even less removed from the standard. Then some particularly privileged blacks, especially after slavery ended and they took their places as leaders of the society, grew up speaking English itself.

Thus the deepest layer of creole on the Guyanese continuum shows the hallmarks of a language developed from a severely reduced English, such as the use of *me* for all first-person pronouns, filled out with African language features such as consonant-vowel structure (*gii* for *give*) and "default" settings such as preverbal particles to indicate tense (here, *bin*). Moving into the semicreole variety, we see *mi* becoming *a,* closer to *I,* the third-person pronoun *am* taking on a shape closer to English *him,* the preverbal particle becoming *did* (which was used in neutral statements in the regional Englishes that blacks heard in the colonial era), and *gii* taking on the final consonant of *give.* Closer to Standard English, the particularly arbitrary aspects of English come in that show that the slaves speaking this variety had all but acquired the language: *give* is past-marked with an irregular vowel change (*gave*), and the pronouns take on their standard forms. *I gave him* in Guyana differs from the same sentence in standard American or British English only in regard to accent (the West Indian "lilt").

Yet in practice there are no sharp divisions between these various levels. On the contrary, they are used in a diglossic context, with most speakers controlling some levels beyond the one that they are most comfortable in. The deepest creole layer is spoken most by people with the least education, especially rural ones, but such people can usually "shift" to a semicreole variety as well. The person for whom a semicreole variety is the home language can usually shift to a standard variety in communication with Standard English speakers and might also shift into deeper varieties to make a joke or underscore a point. Different people control various parts of the continuum, some few able to shift from one end of the range to the other one. But because there are no perceptible "milestones" along the continuum where "creole" becomes "semicreole" becomes Standard English, to a Guyanese, all of these places along the cline are "English." Where outsiders see people speaking "two languages," the speakers themselves simply see there being many different ways one can speak English according to class and social situation.

I will never forget when I boarded an elevator with a Guyanese man (my dissertation adviser) at a conference and another Guyanese man jumped in at the last minute. They started out speaking Standard English, largely in deference to me, but as the elevator went up and their conversation became gradually warmer and more spontaneous, they started gliding into increasingly more creole layers of their speech repertoire. The higher we went, the less of their conversation I could grasp. I lost the first sentence above the fifth floor; by the tenth, all I knew was who they were talking about; by the eighteenth, all I knew was that something was really funny and that it probably wasn't me. By the twenty-fifth, floor, when we got out, they might as well have been speaking Turkish. Yet to them, they had never stopped speaking "English"–they had simply traveled along a continuum of creolized varieties of it leading away from the lone vanilla variety I grew up in. There is a similar creole continuum in Jamaica, and Jamaicans are often perplexed or even discomfitted at the idea that their patois is something other than "English"–and after all, it *is* English, whatever the formal usefulness of professional linguists' taxonomic pigeonholings.

I once took part in a documentary filmed on the Atlantic coast of Nicaragua. Today a Spanish-speaking area, for much of the 1700s this coast was occupied by English settlers and slaves speaking a variety of West Indian patois. The legacy of this era is the unusual one of African-descended people who speak Spanish as a "high," or "public," language but still use the English creole as the familiar "home" language. Unlike in Guyana or Jamaica, however, the creole is not one pole of a continuum extending from Standard English; instead, the imposition of Spanish since the 1800s has lopped off the Standard English end of the cline, such that Standard English is simply one more foreign language to most of the black creoles. Only the few educated creoles are truly comfortable in English, and creole children in particular, not even having yet learned Spanish in school, speak only creole English, just as a child in Stuttgart has not yet learned Standard German and speaks only Schwäbisch.

Standard English, then, is as alien to the kids here as creole English is to most of us, and during my visit I found myself in the unexpected situation of being all but unable to communicate with the children in our respective "Englishes." I had to ask for endless repetitions to grasp even the basics of what they were saying–if you are used to *What's your name?* then a little girl looking up and asking you "WAH yoo nyehm?" can be almost impossible to process (especially when you do not expect such a direct question from a three-year-old–she was also fond of pointing at people in mock anger and shouting "Doo-OWN TAHK!"–that is, "Don't talk!").

Meanwhile, anything I said to them brought blank stares no matter how slowly or clearly I spoke. I will never forget trying to explain to a local boy that we needed him to stop running in front of the camera while we were filming a scene–I simply could not get the concept across; to him I might as well have been speaking Korean, and even Spanish was of no use because he was too young to go to school.

Like Vermouth in a Martini: Whispers of Pidginization

Finally, if for a long period of time a language is spoken as a second language by more people than learn it as a first language, then generation after generation of people acquiring only, say, ninety percent of its structure rather than one hundred percent and passing this variety on to subsequent generations can give the language a kind of "shave," trimming away some of the more arbitrary aspects of the language that nonnative speakers are most likely not to fully master. This can be thought of as a hint of pidginization, leaving the language essentially intact but carrying signs of having been "streamlined" a tad by adult learners in the past.

Swahili is one of hundreds of Bantu languages spoken in sub-Saharan Africa. Before its establishment as an official language in East African countries like Tanzania and Kenya, it was spoken natively only by small communities on the East African coast and on the island of Zanzibar. However, Swahili speakers were central in trade with Arabs to the north and other Africans inland. As a result, from early in the past millennium, Swahili was spoken as often, if not more, as a second language than as a first. This is why today Swahili, though very much akin to other Bantu languages in most details, is somewhat less complex than the others in some indicative respects. Almost all Bantu languages distinguish words with tones as well as sounds, with tone also being central to other aspects of grammar–in some, for example, a tone indicates the subjunctive. Swahili is rare among Bantu languages in having no tone. Because almost all languages descended from Proto-Bantu do have tone, we can assume that at an earlier stage Swahili did, too. It lost it as a result of how difficult a language's tones are to acquire for people speaking languages without it, as well as even for people speaking languages that use tone differently from the one they are learning, as would have been the case for speakers of other Bantu languages learning Swahili before it lost its tones. Swahili also has fewer irregularities than do other Bantu languages; the learner of Swahili is often delighted at how few exceptions to the rules there are compared with, say, the irregular-verb situation in French or Italian. As second language of choice for centuries for much of East Africa as well as for Arab traders, Swahili was bound to be shorn of some of its "benign excess," pidginized "once over lightly."

When a language has a large number of both native and nonnative speakers, the nonnative variety often shows what the language as a whole would look like if nonnative speakers became a majority. Fula is a West African language that is used as a second-language lingua franca by many people in various countries from Senegal all the way down to Cameroon. If the Romance or Germanic genders are a nuisance for nonnatives, Fula is an absolute nightmare. In French, German, Spanish, or Russian, we must cope with two or three classes that nouns can belong to, and in the first two languages, gender membership is only fitfully suggested by the shape or meaning of the word. In Fula, there are as many as sixteen "genders" that a noun can belong to; except that one of the genders is for people (there is no masculine–feminine distinction), they correspond only occasionally and broadly to classes of meaning. For example, the words for *orange, jar,* and *squirrel* all belong to the same gender, and seeking some similarity in the shapes or ends of the words is no help: the respective words are *leemuu, loo,* and *jii.*

Each gender is marked by a different article: *an orange* is *leemuu-**re***, *a man* is *gor-**ko***. But then the article in each gender has three or four variants, and which variant is used with which noun is, again, arbitrary and must simply be learned by rote. *Leemuu-**re*** "an orange" but *loo-**nde*** "a jar"; *gor-**ko*** "a man" but *pul-**lo*** "a Fula person."

It is not exactly surprising that nonnative speakers of Fula do not use these gender markers as nimbly as native speakers, and in fact, among many nonnative speakers, the markers are essentially not used at all, instead replaced by an all-purpose marker used in all of the genders. In one study, the only nonnative speaker recorded as marking an adjective with its "article marker" had lived with a Fula-speaking family for a long period. These nonnatives'

Fula is by no means "pidgin" or "broken" Fula–it is a fluently spoken language that nonetheless has done away with some of the machinery in the grammar hardest to pick up (there is even some slippage and static in the use of these gender markers by *native* Fula speakers!). If nonnatives came to vastly outnumber native speakers, then the Fula of tomorrow would differ from today's native Fula in having been "cleaned up" on this score and would thus appear somewhat "optimized" in comparison with its relatives such as Wolof and Serer (both of Senegal).

In some cases, it is just one dialect of a language that carries the results of this kind of "shaving" by extensive acquisition and use as a second language. Black English in the United States has sometimes been presented as a creole or semicreole, but almost all of the traits brought to bear in such arguments are actually inheritances from nonstandard regional British dialects that slaves were exposed to, in the vein of the Mauritian *zot ti pe ale* case we saw earlier. This includes the "unconjugated" *be* (still heard today in Irish English), absence of plural marking on some nouns (*two cent*), the *done* past (*She done seen him already*), and several other features. Meanwhile, the dialect lacks hallmark African-created creole features such as the running of verbs together or the use of one pronoun form for all cases (that is, "Me go"). Just as Swahili is very much a Bantu language, Black English is fundamentally a dialect of English, right down to maintaining all of the basic irregular verb forms such as *came* and *went* and irregular plurals such as *men* and *mice*. Yet Black English was indeed created in a situation where a great many adult Africans learned English on plantations as a second language, and the dialect carries a few legacies of that history. The absent *to be* in sentences like *She my sister* is one.

Intertwining Just a Little

Pidginization, then, operates on a continuum. Moreover, there is no dividing line between direct intertwining of languages and the lesser and broader degree of mixture in creoles. Full-blown intertwining was impossible on Caribbean plantations, because African languages spoken in Senegal differ from the ones spoken down in Angola as much as English differs from Hungarian, and thus there was no single "African" grammar that English words could be simply plugged into. In other situations, however, the creators of a creole language spoke distinct yet very closely related languages, resulting in a mixture of a kind intermediate between intertwining and creolization.

For example, when Iberian soldiers married local indigenous women in the Philippines starting in the 1500s, a creole Spanish developed, still spoken today. The women spoke various local languages such as Hiligaynon, Cebuano, and Tagalog that differ only about as much as Spanish and Portuguese do. Closely related languages have in common not only broad features like word order, but even a certain number of relatively specific features in a grammar such as particular endings or pronouns–for example, in both Spanish and Italian, *tu* is "you" (spelled *tú* in the former) and *-o* is the first-person-singular ending on verbs. Because of similarly close relationships between the Philippines languages in question, the creole is decorated with a number of prefixes and grammatical words that these local languages had in common, courting the intertwining in Media Lengua.

But only just. The Spanish creole of the Philippines does not include anything approaching the full complement of the local languages' fearsome array of prefixes and suffixes necessary to negotiate the relationships between verbs and the nouns around them, and certainly not their *in*fixes, which we only have in English in the occasional construction such as what has actually been termed by linguists "fucking-insertion," as in *fan-fucking-tastic!* Here is a sentence in the Spanish creole, artificially constructed to contain as many of the Philippines language elements as possible; the words in bold are from those languages:

Hindi'	**kitá**	ay-	**man-**	encuentro	el	**mana**	muher.
not	we	will-	VERBAL-	meet	the	PLURAL	woman

"We will not meet the women."

Yet there are not a great many more Philippines elements than these, and they are incorporated within a general trend of simplification even of the local languages' grammars, manifesting the roots of the creole in what began as a pidgin rather than as a simple combination of Spanish and the native languages unintermediated. The *-man-*, for example, marks verbs under certain conditions; it is used in a way that is a vast simplification of the use of the prefix in Hiligaynon, Cebuano, or Tagalog. This language, then, is a creole that developed somewhat in the direction of direct intertwining, and there are other such cases.

Creoles: Butterflies Out of the Cocoon

Thus although most of the descendants of the first language, regardless of the endless subbranchings upon subbranchings that have developed in the past 150,000 years, trace back uninterruptedly to the original language, there are cases where a subbranch's development among some speakers is interrupted when they acquire only the rudiments of the language's vocabulary and structure. These speakers rebuild the language from the ground up into a new one, which then proceeds to live on like the thousands of other languages and dialects of languages, undergoing the very same processes as they do.

Creoles, like other languages, change through time. In Surinam, many slaves escaped the plantations and founded communities in the forest where their descendants survive today. They speak a descendant of early Sranan called Saramaccan. In early documents of Saramaccan compiled by Moravian missionaries, sentences are negated with *no: Kofi no waka* "Kofi is not walking." Today, however, verbs are negated with a different word, *a: Kofi a waka.*

Because they are generally spoken by several separate subcommunities, creoles, like other languages, branch into dialects. In the more southerly Saramaccan-speaking villages, that negator is *an*, nasalized as "AH^{ng}."

Creoles, like other languages, mix with other languages. Sranan and its offshoot Saramaccan derive most of their core vocabulary from English, because Sranan developed when Surinam was briefly run by the English. But Dutch has been the official language of the country since 1667, and Sranan and Saramaccan are thus full of words from Dutch. The Saramaccan numbers from one through twelve are *wán, tú,* ***dií,*** *fó,* ***feífi,*** *síkísi, sében, áiti, néni* or ***néigi,*** *téni,* ***elúfu, tuwálufu,*** the bold ones being from Dutch rather than English.

Creoles, like other languages, can even be pidginized themselves. Just as the plantation contract system in Oceania created language-learning conditions quite similar to those in Caribbean slave societies, in South Africa when miners speaking various languages were imported to work in mines and were exposed to Afrikaans only in passing exchanges with foremen, the stage was set for a new pidgin to arise. Today, the pidgin, called Fly Taal, has been recruited as an everyday language for many black South Africans. The pronoun *hy* "he" is used for women as well as men; the word order is subject-verb-object instead of placing the verb at the end of some sentences as Afrikaans does; some aspects of Bantu grammar such as the plural prefix *ma-* are used (*ma-gents* means "gentlemen"). Fly Taal, then, is a pidgin of a semicreole, quickly becoming a creole itself.

"There's No Such Thing as 'a Language'" Revisited

The four chapters so far have put us in a position to fully appreciate how much richer the nature of how speech varieties occupy the globe is than the scenario implied by political maps–how multifarious a language's various manifestations are, how languages shade into one another geographically, and how they shade into one another structurally and give birth to new hybrid languages.

On the opposite page, for example, we see what "Spanish" actually consists of, taking into account its origins in change and dialectalization and its subsequent history of mixture and pidginization.

"Spanish" is actually a complex of dialects, for one. "Portuguese" emerged out of what began as a cluster of dialects on the Iberian Peninsula, not considered a separate "language" until the previous millennium. Even today the Galician dialect is a bridge between Spanish and Portuguese, being a dialect of the latter but with much in common with the former. In Uruguay, which abuts Portuguese-speaking Brazil, speakers of Spanish and Portuguese communicate in an intertwining of the two called Fronterizo.

The Latin American Spanish dialect has subdialects in each Latin American country, Uruguayan being one. In Ecuador, the local Spanish dialect intertwines with Quechua as Media Lengua; Quechuans also speak a Spanish variety that, though not intertwined, is influenced by Quechua about as much as Irish English is by Irish. Meanwhile, in the United States, Mexican Spanish has undergone heavy lexical influence from English to become what is

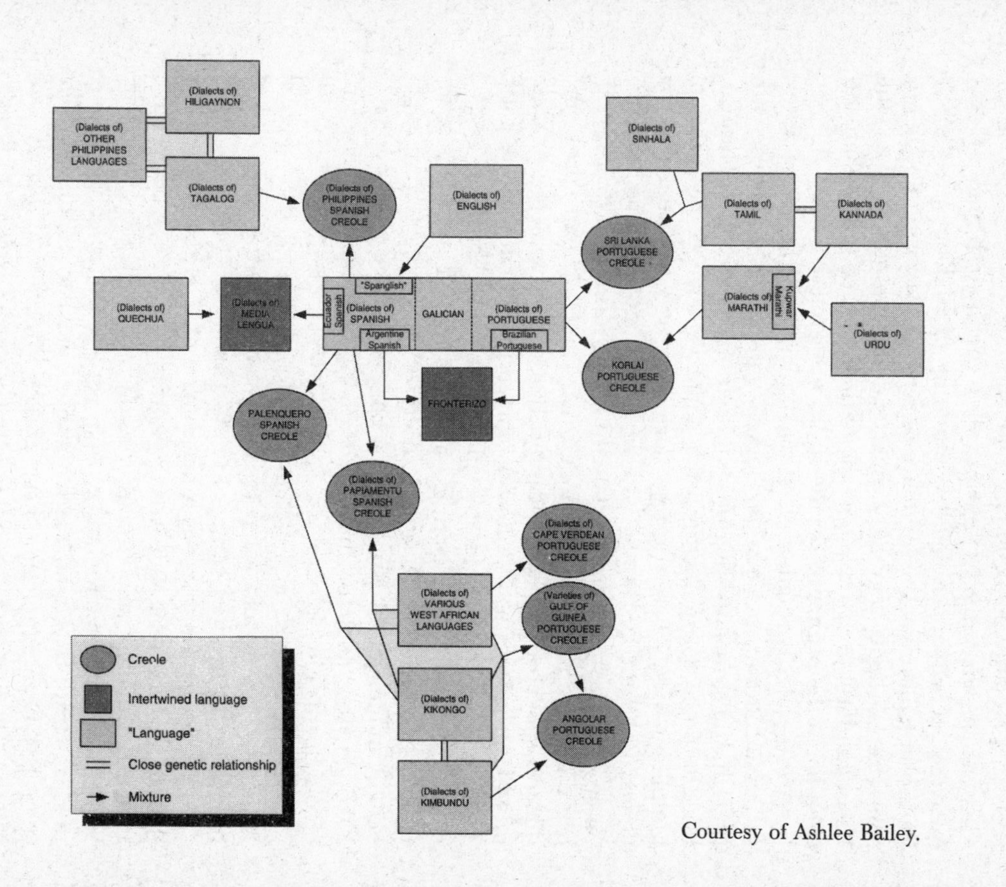

Courtesy of Ashlee Bailey.

popularly known as "Spanglish," a new dialect. In the Philippines, Spanish has become a creole displaying hints of intertwining; the creole itself comprises three or four dialects (depending on where you draw the line) plus one now extinct.

Over in the Western Hemisphere, in the Spanish creole Palenquero spoken in the isolated community of El Palenque de San Basilio in Colombia, the influence of Kikongo is strong enough to constitute a light degree of intertwining: the creole uses the Kikongo plural prefix *ma-* (cognate to the one used in Fly Taal) and even has some pronouns borrowed from Kikongo.

Meanwhile, Portuguese itself has undergone similar processes. In India, Marathi speakers developed a Portuguese creole rife with Marathi traits; meanwhile, Marathi itself, in the village of Kupwar, has altered its grammar toward Kannada's through centuries of multilingualism. There are also Portuguese creoles influenced by Gujarati, Kannada, Tamil, and Malay. Standard Brazilian Portuguese differs from Continental Portuguese more than British English differs from American (though there are Brazilian Portuguese dialects spoken by isolated peasants that are semicreoles, because of widespread use as a second language by Africans in the past). There are a Portuguese creole spoken in Cape Verde (in several dialects) and the dialect complex of others spoken on islands in the Gulf of Guinea. The Angolar variety of this dialect complex is heavily influenced by Kimbundu, a Bantu language quite similar to Kikongo, the language that has similarly influenced Palenquero creole Spanish in Colombia.

Finally, Spanish and Italian are close relatives such that, in Argentina, immigrants speaking the two languages formerly communicated in an intertwining of the two called Cocoliche.

On the opposite page is a diagram, similar to the Iberian one, that shows what "German" really is from a bird's-eye perspective. Dutch is actually a standardized version of what began as northern varieties of German and, until recently, one could travel from village to village from Holland to Germany along a dialect continuum. German has been creolized once, as Unserdeutsch in Papua New Guinea. Dutch gave birth to three creoles, one of which was Berbice Dutch, actually a case of light intertwining like Philippine creole Spanish, with one dialect of the African language Ijo contributing about a quarter of the vocabulary and even some suffixes and grammatical words (such as *en, mo,* the ending *-te, eni,* and the plural

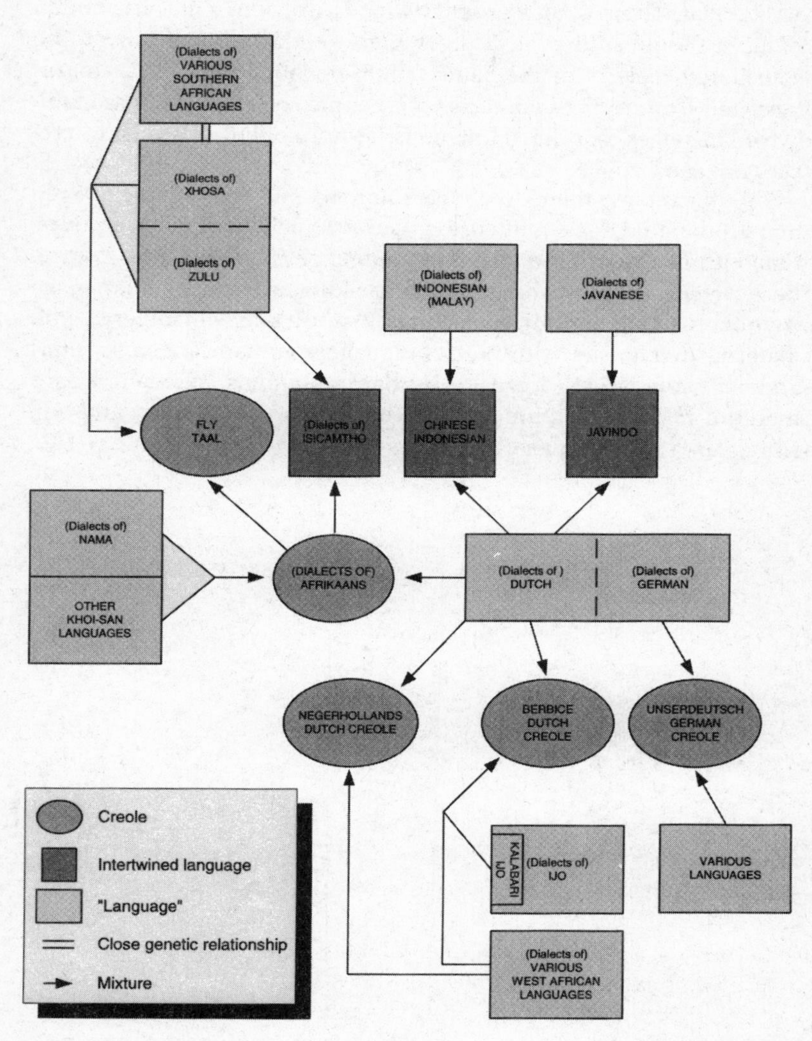

Courtesy of Ashlee Bailey.

ending *-ap* in the sentence on page 159). Dutch intertwined with Javanese in one instance and with Malay in another. Then there is the semicreole Dutch, Afrikaans, which itself has been repidginized as Fly Taal. There is also an intertwined Afrikaans-Zulu, Isicamtho. Zulu, in the meantime, is close enough to Xhosa for the two to be considered dialects of the same language; both have click sounds inherited from ancient contacts with people speaking the same family of languages that the "Hottentots," who pidginized Dutch to create Afrikaans, spoke.

With creoles, then, we have rounded out our perspective on human language as guided by certain ineluctable imperatives. Once it hits the ground, a human language must and will change. Because change can proceed in various directions, once a language is spoken by separate populations, it must and will diverge into dialects. Juxtaposed with other languages, human languages must and will mix. Torn down to its bare essentials, if needed as a medium of full communication, a human language must and will rise again as a new one.

5

The Thousands of Dialects of Thousands of Languages All Developed Far Beyond the Call of Duty

Today when we hear an old standard like "My Funny Valentine" or "I Get a Kick Out of You," it is with a musical accompaniment tailored to the individual performer. But a great many of these songs were first heard within the scores for stage musicals premiering in New York.

In the 1930s and 1940s, a vogue reigned for orchestral arrangements for the performances of these songs that were vastly more elaborate than was strictly necessary. "This Can't Be Love," from Richard Rodgers and Lorenz Hart's score for *The Boys from Syracuse* (1938), is in itself a straightforward little tune, yet for its original performance, orchestrator Hans Spialek served up a marvelous instrumental ambrosia underneath the singers. As the singers put the first sixteen bars over the footlights, strings and woodwinds trip all over each other in filigrees of chromatically descending thirds–writing so beautiful that it would be worth hearing even without vocal accompaniment.

The ironic thing about these touches is that virtually no one in the theater audience ever noticed them, especially in the days before amplification. Besides, even today an audience member is too caught up in watching the performers to pay attention to such things. Moreover, in the 1930s, orchestrators cannot have been hoping that people would glean their artistry during repeated listenings to recordings of the show, because musicals and their songs were almost never recorded with their theater orchestrations until the early 1940s.

It is hard even to make the case that these elaborated orchestrations were intended as artistically integral parts of the songs, even if not consciously perceived by the listener. The woodwind filigrees in "This Can't Be Love," for example, have no perceivable relationship to the lyric. The particular extravagance that the orchestrators of this era lent their craft was, in the end, as utterly unnecessary as the fancy fins on 1950s cars were to their mechanical functioning.

Listening to the lusciously busy instrumental backgrounds in these scores, one may think, "Why in the world did they bother?" The answer is that top-quality orchestrations in this era were considered a "touch of class," rather like the chocolate drop that hotels often leave on your pillow. The world's languages are densely overgrown just like these old orchestrations, with no more necessity to sophisticated human expression than that chocolate drop's necessity to your returning to the hotel in question.

In any language on earth, the structure of the grammar and the subdivisions of meaning it can convey far overshoot what would be necessary to even rich and nuanced communication. All languages are provided with the equipment to convey sine qua non fundamentals of communication. They all have, for example, nouns, verbs, pronouns, words to indicate the position of an object in relation to others (such as *under, on, behind*), etc. However, there are more potential "spaces" that a language can fill, all of which a native speaker of a language might suppose "must" be filled but which in fact could not be, with no detriment to communication.

In English, we distinguish *I found the money* from *I have found the money*; the former implies that the event is past and done, whereas the latter implies that the effects of the finding extend into the present, that the finding of the money will serve as the basis for further developments. We English speakers think of this as quite "normal": we encounter it in Spanish and the form, if not quite the same function, in French, German, and Latin; Ludwig Zamenhof dutifully included it in creating his artificial language Esperanto, his European linguistic roots making it naturally seem to him that even a language designed to be maximally "simple" and "universal" must certainly have a past/perfect distinction.

Yet a great many languages do not make such a distinction overtly and usually just say *I found the money* in both cases, the nuance conveyed by *I have found the money* simply left to context.

At the end of Chekhov's *Seagull*, English translations have Dorn telling Trigorin, "Konstantin Gavrilovich has shot himself," conveying the sense that this event was a recent one with horrific implications for the present. But in the original Russian he simply says that Konstantin "shot himself"–*zastrelilsja*–with no overt indication of the perfect, which would be impossible in Russian. There is certainly no reason to suppose that Russians do not perceive the semantic difference between a past event that is completely past and one that has implications for the present–it's just that their language does not *require* one to mark the distinction overtly, because context almost always makes it clear. Thus, contrary to what we might think having been exposed only to Romance or Germanic languages, overtly indicating the past/perfect distinction is not integral to successful communication: strictly speaking, it is a decoration–useful and even pleasing if it's around, but ultimately nonessential.

In a global sense, this distinction between "past" and "perfect" is one potential space that English and some other languages have grown into marking regularly but that most languages have not. This is one of many such features in English. Meanwhile, other languages on our "language bush" have extended their tendrils into various other spaces, many of which strike us as counterintuitive when we encounter them but which in the end are results of the same process of benign overgrowth that our *have*-perfect is.

The first language, then, split into thousands of branches that each have evolved in part to maintain what is necessary to communication but in equal part have evolved just because various semantic spaces, perceivable to and processible by human cognition but nonessential to the needs of speech, were "there" to be evolved into.

The Incredible Overzealousness of Language

Dotting the *i*'s and Crossing the *t*'s: Evidential Markers

There are many languages in the world where it is an integral part of even basic expression to indicate not just what happened, but your source of the information. In English, we can do this if we need for some reason to be particularly explicit about such things: *He's being courted by Goldman Sachs, so they say; It's coming from inside–at least that's what it sounds like.* But in other languages, notions like *so they say* and *that's what it sounds like* are conveyed by suffixes, which are as necessary to the sentence as marking tense is in English sentences.

In the Tuyuca language spoken by a tribe in the Amazon rain forest, to say that someone is chopping trees requires that one also specify how one knows this. If you *hear* that someone is chopping trees, then you say:

Kiti-gï tii- **gí**.
chopping-he "-ing"

"He is chopping trees" (I hear him doing it).

where *gí* serves to indicate that this is something you heard. But if you actually see him chopping trees, then you say *Kiti-gï tii-**í**,* where the *í* indicates your having seen rather than heard it. If you have not actually perceived him chopping trees but have reason to suppose that he is doing so, then the sentence is *Kiti-gï tii-**hɔi***; if your source of information about the tree chopping is hearsay, then this requires a special marker: *Kiti-gï tii-**yigï**.*

Linguists call these *evidential markers,* and they are commonly found in Amerindian languages, but also in others throughout the world; it's something a linguist is not surprised to find when working out the grammar of a hitherto unanalyzed language. Obviously, speakers of languages without such markers perceive the semantic nuances in question, given that they can specify source of information if need be. Just as obviously, to have markers of such nuances renders a language more expressive and precise in this area than one without.

I remember being at a party in college where one woman came upstairs and said of a particular couple, "They're making out," much to the dismay of another woman who would have preferred that the male half of the couple in question would make out with her. The woman who had made the announcement then floated away into another room, and the disappointed woman spent the next ten minutes running around ascertaining whether what the other woman had said was true, how she could have known it, etc. Did someone tell her they were making out? Did she actually see it, in which case was the couple really doing this in full view of the various other people downstairs? If this party had taken place among the Tuyuca, *They're making out* would have had to be accompanied by an evidential marker, which would have immediately let us all know what the source of the observation was.

When they do happen to emerge, evidentials develop through the same kind of grammaticalization as other suffixes do, beginning as free words and gradually becoming appendages. Here is an example from the Native American language Makah, spoken by the Pacific Northwestern tribe recently notorious for reclaiming its whale-hunting tradition. In Makah, "it's bad weather" translates as a single word, *wikicaxaw*, meaning roughly "it bad-weathers." However, an array of suffixes allow speakers to indicate just where they got the information from. When a suffix is added, the final *-w* becomes a *-k* (languages are needlessly complex!) and, with that in mind, here are three of the evidential endings:

> *wikicaxak-**pid***
>
> "It's bad weather–from what it looks like." ("Looks like bad weather.")
>
> *wikicaxak-**qad'i***
>
> "It's bad weather–from what I hear." ("Sounds like bad weather.")
>
> *wikicaxak-**wad***
>
> "It's bad weather–from what they tell me." ("They say it's bad out.")

Each of these endings began as a separate verb. The *-pid* originally was "it is seen," the *-wad* came from a verb that meant "it is said," and so on. Now, they are kludged onto the ends of other verbs, and sound change through the ages has rendered them quite different from the regular verbs they started out as.

Yet at the end of the day, a language does not "need" evidential markers anymore than a car needs cup holders. It's nice if they're around, but no big deal if they aren't. Evidential markers are an "accessory" to full human communication: certainly English is not "impoverished" or "primitive" in lacking them, for example. English just happens not to have evolved its way into this particular cranny of semantic space.

"Well, It Depends on What *Have* Means": Inalienable Possession Marking

"Having" is a richer concept than a language such as English often leads us to consider. You say that you "have" a head and that you "have" a couch, but really, the former kind of having is a brand of integral composition, whereas the latter is a matter of possession. Your relationship to your head is tighter than your relationship to your couch, presumably, at least–you cannot remove your head, you would not be yourself without your head; the couch, on the other hand, could be separated from you with no physical effects.

Did you ever think about this? I sure didn't until I became a linguist, but this difference is quite plain to speakers of many languages, in which how "having" is marked in the grammar differs according to how integral, or, as linguists put it, *inalienable,* the possession is. In such languages, to express the possession of a body part or family member (in most cases our relationship to our families is of a higher order than that to our furniture) requires a construction different from the one you use to express proprietary possession of an object. In Mandinka, *your father* is:

i faamaa
you father

where "your" is expressed simply with the pronoun *you*. But *your well* is:

i la koloŋo
you 's well

where one must use a marker roughly equivalent to our *of* or the Romance *de.*

Romance and Germanic languages (other than English) have a whisper of this distinction when it comes to actions on body parts: in Spanish, to say *He grabbed my hand,* you say *Él me agarró la mano* "He grabbed to me the hand," whereas *He grabbed my broom* is, more intuitively for English speakers, *Él agarró mi escoba*. Yet in general, the *inalienable* versus *alienable* possession distinction is, by chance, a route that the Berlitz and First World languages have not taken.

Once again, the feature is nice to have around. In Navajo, one can use it to distinguish "her breast milk" from "her milk from the store": ***bi**-be'* is the "integrally possessed" milk, whereas ***be-' a-be'*** is milk possessed like a couch. Yet none of us who speak languages without this kind of marking have ever found ourselves confused over what sort of possession was implied when potentially ambiguous cases like this come up. A language that evolves such an overt distinction does so as a result of random drift.

This can occur in a language that at first expresses both kinds of possession in one simple way: saying *I book* for *my book, you book* for *your book*, etc. (quite common–Chinese speakers will be familiar with it). Through time, an expression arises in which one designates some things that one possesses as being "on" or "at" one, such as the Mandinka example *i la koloŋo* "you-at well"; that is, the well that is at you. This expression is often used only with external possessions, because it is redundant to refer to one's arm as being "on" or "at" one, nor does one (at least usually) sense one's mother as being a convenient accessory to one's being–your mother is more "in" you than "on" you. Thus after the "at me" expression has set in for bags, fruit, and real estate, one continues to say *I arm, you mother*. The result is a language where intimate possession is marked differently from proprietary possession.

To someone who grew up with such a language, it initially seems odd that other languages do not mark this distinction, and a Ludwig Zamenhof who grew up speaking Navajo or Mandinka would most likely have made sure that even the simplest possible artificial language had a different way of marking possession of an ear from that of marking possession of an earring. Only from a cross-linguistic perspective does this, like so much else in all languages, reveal itself as a jolly overgrowth.

Linguistic Narcissism: Inherent Reflexivity Marking

Another area where a language can take a ball and run with it is reflexivity. All languages have a way of expressing that an action is done to oneself, such as English's *I bathe myself.* Yet in English, we are only required to mark this for purposes of explicitness and can quite often leave reflexivity unmarked with no resultant ambiguity–we can say *I bathe* without fearing that someone will wonder, "Well, actually, *whom* do you bathe?" But in many other languages, one must mark the reflexivity overtly in all relevant cases: in French, one must say *Je **me** lave*–*Je lave* implies that you bathe someone other than yourself; in German, you say *Ich wasche* **mich**, not just *Ich wasche.*

Furthermore, in most such languages, the habit has extended to all actions entailing exertion on oneself, most of which an English speaker does not even conceive of as reflexive: in French, one does not just slip, but slips oneself–*je **me** glisse*; in Spanish, one does not just sit down, but sits oneself down–*yo **me** siento*; in Russian, one did not *tire*, but tired oneself–*ja utomil**sja***. This is where the "changes itself" aspect of such languages comes in. In French, one changes *oneself* into a swan rather than just changing into it: *Je **me** transforme en cygne.*

Yet getting a "feel" for such a language entails wrapping one's head around the fact that this fixation on marking any hint of exertion on oneself has spread even into actions exerted *within*, rather than *on*, oneself. In French, then, one does not just faint, but "faints oneself" (*Je **m'**évanouis*); in Spanish, one does not just feel happy, but "feels oneself happy" (*Yo **me** siento feliz*); in German, one does not just remember, but "remembers oneself" (*Ich erinnere **mich***)–similarly. This can even apply to nonsentient objects: in Spanish, if a window broke, it "broke itself"–after all, it didn't break something else, and thus ***Se** quebró la ventana* "The window broke."

Languages like this have developed a "tic" leading them to sniff out and mark any whiff of reflexivity in an action. As with evidential markers and alienable possessive marking, taking reflexive marking this far serves no "purpose"–a language marking reflexivity only in the most literal cases such as *He hurt himself* supports not only fluent and precise communication but noble literatures. As such, marking the "technical" reflexivity of such things as becoming sad and balloons popping is a feature found in only a minority of the world's languages: once we get beyond Europe, in particular, this trait, called inherent reflexivity marking, is more likely to be absent than not. Reflexivity fixation, a sort of linguistic narcissism, is an embellishment on what is necessary for humans to communicate.

The Unexpected Superfluousness of Your *A*-ness: Definite and Indefinite Articles

Nothing seems more basic to an Anglophone than distinguishing explicitly and consistently between *the* man that we were talking about and *a* man that comes up for the first time in a conversation, especially because the languages we learn most often have similar words with similar meanings. Yet, actually, a great many languages do without articles. All languages have some way of distinguishing the definite from the indefinite, but most do not do so *overtly* as consistently as English.

Russian, for instance, has no articles at all. One can convey that a subject is indefinite–that is, new to the conversation–by placing it at the end of a sentence: *Poezd prišjol* is "The train came," whereas to indicate that *a* train came one puts the word for *train* at the end and says *Prišjol poezd.* This is the kind of thing that led one child of a Russian immigrant I met to tell me that Russian "has no word order," but actually the variations in order that look odd or random to an English speaker convey the same thing that we convey with our articles. Yet this only works within the limited domain of simple utterances like *Prišjol poezd.* There is no way in Russian to distinguish between all of the definiteness distinctions between the English *The solution turned out to be to fashion a bar of soap into the turtlelike shape she had found so charming on the trip* and *A solution turned out to be to fashion the bar of soap into a turtlelike shape she had found so charming on a trip.* But there is no need for such distinctions in grand view, anymore than we have a need to say "I tired myself" because Russians say *Ja utomilsja.*

Indeed, English is rather odd in having words to convey both *the*-ness and *a*-itude[1] on every singular count noun in any sentence: by one estimation, only one in five of the world's languages have words, prefixes, or suffixes to convey both definiteness and indefiniteness. As often as not, a word that translates as *that* is used to convey definiteness when explicitness is needed (*that movie I saw last night*), whereas indefiniteness is not marked at all (Japanese is of this type); in other cases, word order can convey the distinction if necessary. The development of obligatory definite and indefinite articles is, then, another example of a language spreading its tentacles into spaces it need not. The articles in Romance developed from words that meant *that* and *one* in Latin by grammaticalization, where the originally explicit meaning of *that* gradually faded into the less vivid one of *the*, whereas the "one particular" or "singular sensation" meaning of *one* faded into the faint echo of this meaning that *a* entails.

---

1. Once was enough.

Obsessive-Compulsive Pigeonholing: Gender Markers

Another example of how a language overshoots the mark of functionality is the gender markers that are such a pain for English speakers to master in European languages. Although evidential markers, alienable possessive markers, inherent reflexive marking, and articles all give overt expression to concepts that all humans perceive even if their languages do not mark, gender markers in languages such as French, Spanish, German, and Russian stand out in that they generally do not express any real-world concept. Pure ornament, they just sit there. Of course, these markers at times distinguish a biologically male entity from a female one: French *chat* versus *chatte* "female cat." But when we come to the usual cases, like the Spanish *un oj**o*** "an eye" versus *una mes**a*** "a table," we have passed into the realm of arbitrariness, resulting in the seemingly deliberately confounding trio in German, *die Gabel, der Löffel,* and *das Messer* for *fork, spoon,* and *knife.* English does just fine without gender marking, as do a great many languages worldwide. These excrescences develop accidentally through time by language-change processes.

"But how?" one might ask–why would people suddenly start designating eyes as male and tables as female? The process is more easily gleanable in languages with many more genders than European languages' usual two or three. Swahili, for instance, divides its nouns among seven "genders," each with a singular and plural marker. The genders correspond in rule-of-thumb fashion to categorizations of type. One gender contains people, marked with an *m-* prefix: *mtu* "man," *mtoto* "child." Another with an *n-* prefix contains animals: *ndege* "bird," *nzige* "locust." Yet another contains abstract nouns: *umoja* "unity," *ukubwa* "size." This arose because the prefixes began as separate words, the *m-* as some word meaning *person,* the *u-* as a word meaning something along the lines of *realm.* Thus *mtoto* "child" would have originated as two words "person child," whereas *utoto* "childhood" would have arisen as "realm child"–that is, "childness." In time, pairings of this kind were rebracketed into a single one: the general class word was reduced to a mere prefix. After a long period of change, the result was a language where each noun belonged to a "class" marked by a prefix–

a gender system to tax all second-language learners henceforth (other than those speaking other languages in Swahili's Bantu subfamily).[2]

There are even languages where nouns and their gender markers are still in the stage intermediate between being separate words and being one single word. In Cantonese, for instance, in many contexts one must preface a noun with one of several dozen words designating what type of object it is, called classifiers. The word *jēung* is used with flat, horizontal objects, and thus *the table* is *jēung tói*, *the ticket* is *jēung fei*, *the sheet of paper* is *jēung jí*. *Gauh* is used for lumpy things: *gauh sehk* "the stone," *gauh chaatgaau* "eraser." *Ji* is for cylindrical things: *the pen* is *ji bat, the flute* is *ji dék*.[3] These classifiers are the beginning of what could, millennia from now, become fused gender prefixes as the processes of grammaticalization and rebracketing continue apace, and they parallel what Swahili (or more properly, a distant ancestor of the Proto-Bantu language that gave rise to Swahili and its hundreds of Bantu relatives) would have looked like in days of yore.

French's distinction between *le livre* and *la lune,* then, traces all the way back to Proto-Indo-European and beyond, most likely beginning as a long-extinct tribe's cosmological division of nouns into masculine, feminine, and neuter classes based on folk conceptions forever lost in time (some descendants of the language this group spoke, like German and Russian, retain all three classes, whereas many others including French have collapsed them into two). Even when these distinctions corresponded to real-world conceptions of speakers, this was incidental to the essential requirements of a language. After all, most humans lead full lives without making sure to always refer to the moon as a lady or genuflecting to the manliness of their feet. The finer-grained distinctions in Cantonese and Swahili are just as much add-ons to a language: the

2. Actually, because no language behaves perfectly, the situation is not quite this tidy: in some genders, the prefix is more likely to be absent than to actually appear, stray nouns having been tossed into the gender through time.

3. This is an example of how a language can distinguish *the*-ness without having two articles, one definite and one indefinite as English does: in Cantonese a classifier roughly renders a word definite–*gauh sehk* is "the stone"; *sehk* alone would be "a stone," as in "just some stone" or "stones in general."

flatness and horizontalness of a table is too obvious and exceptionless to require regular address (concave tables set perpendicularly to the ground are a tough find). Even at its outset, then, gender marking indicates a certain benignly obsessive-compulsive corner of linguistic habit. Where there is no longer any perceptible link between the marking and the world as its speakers now conceive it (male forks, female spoons, and knives that never develop the urge), we are faced with a sterling demonstration of how the human mind can handle a great deal of linguistic sludge superimposed on what a grammar is really for.

Testing the Limits: How Much Can Humans Handle?

For my money, there are few better examples than Fula of West Africa of how astoundingly baroque, arbitrary, and utterly useless to communication a language's grammar can become over the millennia and yet still be passed on intact to innocent children. Fula has as many as sixteen "genders" in the sense that we parse the concept in Indo-European languages, and which gender a noun belongs to is only roughly predictable beyond the one gender that contains humans. Moreover, within each gender, the marker varies arbitrarily according to the noun: *leemuu-**re*** "orange" but in the same gender class is *loo-**nde*** "jar." Adjectives, instead of taking a "copy" of the marker variant its noun takes, take their own particular marker variant, which must be learned with the adjective. Thus a big orange is not *leemuu-**re** mau-**re***, but *leemuu-**re** mau-**nde***.

Fula is even more elaborated than this. There are often not just two, or even three, but four variants of the gender marker. Thus in our "gender" that contains oranges and jars, the marker turns up as *-re* in *leemuu-**re*** "orange," as *-nde* in *loo-**nde*** "jar," but as *-de* in *tummu-**de*** "calabash"; in another gender, a strip is *lepp-**ol***, a feather is *lilli-**wol***, a belt is *taador-**gol***, whereas a leather armlet is *boor-**ngol***. Besides this, like any language, Fula has its irregulars: in one gender, you have to know not only that a noun will take either *-u, -wu, -gu,* or *-ngu* as its marker, but also that the occasional noun will go its own way and take *-ku* instead.

On top of all this, Fula also has an additional complexification, of a sort found only occasionally among the world's languages. One of its genders is for diminutive versions of things, such as *little boy* versus *boy*, and another is for the opposite, augmentative versions, such as *big rock* or *boulder* versus just *rock*. By giving a "neutral" Fula word the marker for one of these classes, we create a diminutive or augmentative version of that noun–thus, there is a diminutive "gender" and an augmentative "gender" (talk about the male and female genders just being societal constructs!). That alone is no big deal, but Fula can't just leave it at that (the way its distantly related Bantu cousins such as Swahili do).

To wit: for most Fula nouns, when we tack a new diminutive or augmentative gender marker onto the *end* of the word, simultaneously the consonant at the *beginning* of the word changes in some way–and as often as not there are two different consonants that it might become, depending on which of the genders we are switching to. *A man* is *gor-**ko***. The article for the augmentative "gender" is *-ga* (this is actually one of four variants for this gender plus an additional one that pops up irregularly!), but one says not *gor-**ga*** but ***ngor-ga***. There is another consonant change to make *gor* plural: the plural article (like French *les* or Spanish *los/las*) is *be*, but one says not *gor-**be*** but ***wor-be***.

All that is regular in this process, then, is that the consonant will change if the noun is given an article for diminutive, augmentative, or plural. But each consonant makes its own particular changes; a monkey, *waa-**ndu***; some monkeys, ***baa-di*** (*di* rather than *gor*'s *be* because the singular genders each have their corresponding plural ones!); big monkey, ***mbaa-nga***.[4] And then some consonants have only one other consonant that they change into across the board instead of two; some consonants just don't change at all; and finally there are many nouns beginning with consonants that just refuse to change on *these* words even though they do on others; as with so much in Fula, you just have to know.

And finally back to those adjectives: not only does an adjective take its own gender marker instead of copycatting its noun, but an adjective also undergoes its own consonant mutation according to whatever its first consonant is. If you thought the occasional little

4. There is a separate gender conveying not just "little" but what translates perfectly as nothing other than "shitty little": shitty little monkey: *baa-ngum*. (The augmentative conveys not just "big" but exactly what is conveyed in modern colloquial American English as "big ol' "–big in a slightly clumsy but possibly endearing way.)

ripple in Spanish like *la mano blanca* "the white hand" was bad, where masculine-suffixed *mano* is counterintuitively followed by a feminine-marked adjective, then imagine having to make your way in Fula where "a living monkey" is *waa-**ndu** yeet-**uru***, but "a *little* living monkey," because of consonant mutations proceeding independently in the noun and the adjective, is ***baa-ngel geet-el***, and this sort of thing is par for the course, central to almost any sustained communication in the language![5]

Language and Functionality: An Imperfect Fit

Like the other aspects of language change, the benign overgrowth of language has an analogue in the evolution of animals and plants. Not every aspect of a life form is an adaptative advantage created by natural selection. As Richard Lewontin puts it, many features are due to:

> . . . random fixation of nonadaptive or even of anti-adaptive traits because of limitations of population size and the colonization of new areas by small numbers of founders; the acquisition of traits because the genes influencing them are dragged along on the same chromosome as some totally unrelated gene that is being selected; and developmental side effects of genes that have been selected for some quite different reason.

The analogy here is not perfect. There are indeed equivalents to all three of the processes Lewontin mentions in how language changes, but these processes do not happen to entail the grammar's oozing into the overt expression of *new* semantic categories. Nevertheless, Lewontin's observation is valuable here because it illuminates that not everything about a living system is adaptive in nature. Certainly such systems owe their existence to an evolutionary process driven by adaptive mechanisms, but their development entails the creation and dragging along of a great deal of what linguist Roger Lass calls "junk" in the process. This points up the fact that the grammatical overgrowth we are faced with emerges largely

5. There is even more bric-a-brac hiding in the woodwork: the article variant in *yeet-uru* is really just *-ru*; there is a rule adding the extra *u* after a consonant like the *-t* at the end of *yeet*.

independently of "cultural" factors: our point here is not that evidential markers, gender agreement of the ***la casa blanca*** or ***baa-ngel geet-el*** variety, or the dutiful marking of a window breaking as "breaking itself" are manifestations of the cultures of the languages' speakers. Certainly culture can create new structures in a grammar–it's no accident that societies traditionally as rigidly hierarchical as Japan and Korea have developed different words for eating and giving depending on the rank of whom one is speaking to (elder, teacher, intimate, etc.), and (especially in Korean) special endings corresponding to these status considerations. (Nor is it surprising that among children of Japanese and Korean immigrants to America, growing up in a much less overtly hierarchical society, even when they speak the language, aspects like these are among the most likely not to be learned well.)

However, the bulk of any grammar of a language does not lend itself to this kind of analysis–the features usually evolved simply as the result of the natural overzealousness of language change. One might be tempted to suppose that the Tuyuca-speaking Amazonians developed evidential markers because of a cultural predilection toward close attention to sources of information. But the problem is that evidential markers' worldwide distribution is relatively random: one language will have them, whereas another one a few inches over on the map will not, with no indication that one culture is warier of truth conditions than another. For instance, intuition, and the modern educated person's dutiful impulse–appropriate in itself–to avoid parsing preliterate cultures as mentally unsophisticated, might lead to a hypothesis that people living in a rain forest might benefit from carefully indexing the source and reliability of all new information. This, though, would leave unexplained why Turkish marks evidentiality as well, despite being the vehicle of a literate and urbane civilization that once dominated a great swath of Eurasia, or why a great many languages spoken by people in challenging environments lack evidential markers.

If we would analyze any cultures as most obsessed with possession, they would probably be First World ones, rooted in notoriously focused conceptions of personal and private property. Yet the overt distinction of alienable from inalienable possession in grammar is by and large a "National Geographic" affair, more likely to be found in a little-known language spoken by groups living on the land than in "tall building" cultures. It would be fun, but ultimately rather ad hoc, to figure out how Germans are somehow culturally predisposed to express themselves as "sitting themselves" and "angering themselves" or to marking a car as a man, a rehearsal as a woman, and a sheep as devoid of sex–but whatever we came up with would also have to account for the fact that French, Italian, Russian, and Polish peoples' languages do the same kinds of things.

The impending death of so many languages leads many linguists to point out the fact that languages are cultural repositories; furthermore, our Zeitgeist cherishes cultural diversity. To say that both of these ideological trends are welcome is an understatement–but neither belie the fundamental chance factor in much of what a language has developed beyond the strict necessities of talking about and coping with the world around us. One theorist parses nonadaptive evolution as mere "background noise against the main evolutionary tune," but Richard Lewontin, calling attention to the sheer weight of what in an organism cannot be traced to adaptive pressure per se, suggests that this biologist "is not accustomed to counterpoint." Similarly, cultural developments acknowledged, a true understanding of how our tens of thousands of dialects of six thousand languages came to be the way they are requires an awareness of how very much in a human language is there just because language change does not stop at the boundary of necessity. Instead, like my cat, language blithely noses into any number of areas just because it can.

A Little Bit of "Dammit!" in Every Language and How It Got That Way

Because the world's languages all branched off from the first one so many millennia ago, all of them have accreted this kind of "sludge," almost always to a considerable extent. It is for this reason that there are no "easy" languages: in learning any language on earth, we come to a point, often right at the outset but always at least after a while, when we encounter something that from the perspective of our own language makes us think "Dammit! How can anybody speak this every day?"

Where Do Prefixes and Suffixes Come From?

This first strikes the English-speaking student of European languages at the discovery, somewhere about the second week, that each noun must be learned with its gender: ***le** soleil*, ***Das** Boot*. Then there are the lists of endings for person and number, different ones for present, often two kinds of past ("preterite" and "imperfect"), future, conditional, imperative, present subjunctive, and sometimes imperfect and even future subjunctive–and on top of this, three classes of verb depending on ending (Spanish *hablar, comer, vivir*), each with its own versions of all of these endings lists. Gender and conjugational endings together are called *inflections*.

The verb endings in particular seem, if overabundant, "natural" to an English speaker because we have some ourselves–but the very fact that we do with so few shows that the plethora of verb endings in languages such as Spanish and Russian are not strictly necessary to language in general. One could communicate in Spanish just as well by saying *yo hablar español* as *yo hablo español*; *du sprechen Deutsch* gets the same point across as *du sprichst Deutsch*. In short, inflections of this kind do not *mean* anything. And certainly the noisome irregular verbs (take German's *sprichst* instead of the "expected" *sprechst*) are not necessary to making oneself understood, nor are they in any sense "cultural"–they are just cracks in the pavement exerted by millennia of trampling and sunshine.

Gender and conjugational inflections develop not because they are "needed," but through gradual, creeping grammaticalization of what begin as separate words. In Latin, one way of saying *I will love* was to say *amāre habeō*, which literally meant "I have to love" but did not convey the sense of obligation that the equivalent expression happens to in English. Instead, it meant that loving is something I "have" on my plate and thus will do in the future. In time, rapid speech and heavy use eroded the *habeō* down and rebracketed the two words into one. The result was a new ending, now a piece of grammar rather than an independent word, today visible in Latin's descendant Italian in *amerò* "I will love," where the *-ò* is all that is left of *habeō*. This same process took place in all six person/number combinations and, with various further sound changes, resulted in a paradigm of future endings in Italian and other Romance languages:

amāre habeō → amerò	"I will love"
amāre habēs → amerai	"You will love"
amāre habet → amerà	"He (or she) will love"
amāre habēmus → ameremo	"We will love"
amāre habētis → amerete	"You will love"
amāre habent → ameranno	"They will love"

This process is the source of almost all of the inflections in any language.

Thus inflections develop essentially as the result of an accident due to wear and tear. They come, they go. Some languages have more than others. Meanwhile, the people go on speaking, whatever the current inflection situation in their language happens to be. All of this shows that, marvelous and exasperating as they are, inflections are incidental to communication. Any language that has them has, in that regard, overshot necessity.

Setting the Tones

Nothing makes the ultimate superfluity of inflections clearer than the fact that, though in some languages they come and go or linger at a low level, in quite a few there are none at all. Chinese languages, for instance, present no lists of endings to the learner and, on this score, appear to be "simple" languages. Yet Chinese is in fact harder for an English speaker than any European language, because the same syllable can have a great many different meanings depending merely on what tone it is uttered.

In Cantonese, for example, there are by most counts six tones, such that you can have a dazzling range of meanings, especially taking into account that there are homonyms in this as in any language. Here are all the things that the syllable *yau* can mean:

high and level tone:	*worry* (or, when combined with other words, *rest*)
high and rising tone:	*paint*
middle and level tone:	*thin*
low and falling tone:	*oil* and *swim*
low and rising tone:	*have* and *friend*
low and level tone:	*again* and *right* (as in *hand*)

Fan can mean *divide, powder, advise, grave, excited,* or *share;* and so on. This is the language that most Chinatown residents are speaking effortlessly and that presents massive challenges to the European unaccustomed to linking meaning to subtle tonal gradations.

The Chinese languages are hardly unusual in this regard. Usually, if a language does not present the stumbling block of lists of endings (and the exceptions to their use), then it presents the alternate challenge of being tonal. This is true of languages of Southeast Asia (Thai, Vietnamese), a great many African languages, and other languages such as various ones spoken by Indians in Mexico. At a party, a woman who had been to Ghana told me that Twi was easy because "it had no grammar." She meant that there was no *amo, amas, amat* to bother with–but Twi is a tonal language with a game plan similar to that of Chinese, and learning it, or relatives like Yoruba and Igbo, requires reorienting oneself toward listening for, and producing, tone as a determinant of basic meaning, something best learned in the cradle. Old travel memoirs by European businessmen and explorers on the Ghanaian coast are full of references to how hard Europeans found it to master the local Twi, which they had assumed was so "primitive"–even when they had often learned European languages other than their own with little comment.

Like inflections, tones emerge in a language as an accident of sound erosion rather than out of any communicative imperative. Specifically, a language becomes tonal as a kind of "desperate measure" when sound erosion has left this as the only way to distinguish one word from another. In Vietnamese of roughly two millennia ago, for instance, words were distinguished by sounds as we are accustomed to. Here are hypothetical examples but with the types of sounds actually used in such processes:

da	"big toe"
da'	"nostalgic"
dah	"shitty little monkey"

The ' indicates the catch in the throat that *Uh-oh!* begins with; the *h* would have been an actual *h* sound pronounced lightly (say *Holly* without the *-olly*), rather than simply indicating the pronunciation of the *a* as "ah" the way we would read it in English.

As time went by, that Old Devil Sound Change wore off the final *h* and the ', leaving just *da, da,* and *da*. However, the three *da*'s did not sound precisely alike. Say *da'* with that "catch in the throat" and notice that your *a* is likely to come out on a slightly lower note than if you just said *da*. Meanwhile, the *h* in *dah* had a way of raising the "note" of the *a* slightly above the regular *da* level (this is less intuitive for an English speaker because we don't have final *h* sounds). Thus before the consonants had eroded, not only they, but technically also these tonal differences, distinguished the words. Here, ` over a vowel means "low tone" and ´ over a vowel means "high tone":

da "big toe"
dà' "nostalgic"
dáh "shitty little monkey"

Of course, these subtle distinctions in tone were essentially background noise in the language when the *h* and the ' were still there to distinguish the words; to not pronounce the tones would simply create a cute but comprehensible "accent." However, when these final sounds eroded, they left behind the tone differences on the vowels before them. At this point, for better or for worse, the tones were all that was left to mark the distinctions in meaning, and human perception is capable of attending to shadings this subtle closely enough to form the very foundation of a language spoken all one's life.[6] As a result, in a tone language like Vietnamese these words are distinguished by the tones alone:

da "big toe"
dà "nostalgic"
dá "shitty little monkey"

One might wonder why they didn't just make do with good old fashioned homonymy, relying on context to distinguish meanings (after all, how likely is it ever to be unclear whether one is referring to one's big toe or one's dog?–"Oh, I'm sorry, Tranh, I thought you were worried that your *toe* was pregnant!"). However, I have given a simplified presentation–there are actually six tones in Vietnamese as in Cantonese, and not just some but all words are monosyllabic. Even the common Vietnamese name *Nguyen* is actually one syllable: the *ng* is pronounced not as "n-g" but as the single sound of the

6. If you grew up with Chinese, Vietnamese, Thai, Twi, or Yoruba, I know this seems mundane. But as someone raised on English and having had the natural Western experience of encountering only languages such as French, Spanish, and German during my formative years, I find this absolutely amazing, rather like a Native American identifying an animal and how long ago it passed by a footprint.

ng in *singer* (think about it–you don't say "sin-gurr"), and the *y* is a vowel-like sound such that there is a wild sploosh of three vowels in the middle, the *uy* being a kind of runway lead-in to the *e*. Thus the word is not the "nn-GOO-yen" that we hear over the PA system at an airport, but something like "^{n}g^{uy}EN." There are only so many consonant–vowel combinations possible in any language,[7] and for a language to be founded entirely on juggling six-way homonymies of the limited possible collection of such little monosyllables would be awkward and confusing to say the least.

Thus the tones end up taking up the slack. Unlike the purely "accessory" features we have seen such as evidentials or inflection, the tones are certainly necessary within the languages in which they exist *such as they have evolved,* making such heavy use of simple monosyllables that tone is thrust to the forefront. Whereas Spanish without inherent reflexivity marking would just be a less metaphysically oriented Spanish where a window broke instead of breaking itself, Vietnamese without tones would be a soup. But–the tones arose in a language that originally neither had nor needed them, because final consonants did the job of distinguishing words just as they do in English in cases like *pop, pot,* and *pod*. A language like Vietnamese now relies on a much finer grained variety of perception to distinguish meaning–more challenging to any second-language learner (other than people speaking languages closely related to the tonal language in question) than perceptions based on consonants and vowels–because of the accumulation of historical contingencies through time.

Tone, then, is not a necessary feature of a human language–it is a cognitively parsable, but ultimately accidental, permutation of a language's original material, which can result only from a language that began without it. There was a queer galumph of a movie in 1984 called *No Small Affair,* in which Demi Moore aspires to be a rock singer and begins specializing in rock versions of old standards as lovelorn Jon Cryer stands by egging her on (needless to

7. Actually, Vietnamese words can end in a few consonants, but this still leaves the potential inventory of words much lower than in a language that allowed, say, *bib, bic, bid, Biff, big, bill, bin, bit,* and *Bix*. And many other tone languages allow fewer final consonants than Vietnamese: Mandarin Chinese allows only *n* and *ng*.

say, the chemistry between these two did not exactly bring Bogart and Bergman to mind). She sings a version of "My Funny Valentine" with chords so greatly altered from the 1937 originals that it takes a musician to even be able to hear how they are derived, through myriad inversions and reinterpretations, from the pat, conventional ones in the original sheet music. Derived from the original chords they are, however, and few pop or rock musicians would write a sequence of chords like these simply off the top of their heads–such an arrangement virtually presupposes the existence of a template from which the new version was derived. Tone is "derived" from a language that relies on sounds to distinguish words rather as Moore's "out there" version of the song is derived from its source, just as one cannot fully "get" my favorite TV show of all time, *Married . . . With Children,* without reference to the sappy family sitcoms that it parodies.[8]

Into Every Language a Little Sludge Must Fall

A very few language groups in the world make do largely without inflection *or* tone, but even here one is confronted with "Dammits" eventually, because any language that has existed for millennia cannot help but have wound its way into something of the kind out of sheer developmental exuberance.

The Polynesian languages, such as Maori, Hawaiian, and Samoan, initially surprise one in that there is neither anything of the *hablo, hablas, habla* variety nor *da, dà, dá*. But alas, these lan-

8. Germans adore the dubbed version, *Eine Schrecklich Nette Familie,* but I don't think they really "get it" the way we do; they just see it as a parody of "America" rather than our old TV shows, most of which they have never seen. Similarly, their dubbing of *South Park* is witless by our standards because the show is so profoundly rooted in subtly ingrained aspects of our particular popular culture and our layered relationship to it. They have the boys croaking along in barely disguised adult male voices; only when their version of Isaac Hayes says, "Hallo, Kinder" in a dead-on imitation of Hayes himself does it hit home (and they even dub Chef's songs into German!). One joke they do carry over nicely is in the episode in which Tina Yothers (of the late sitcom *Family Ties*) comes to town; in the German version, they manage to substitute an American-television second banana unheard from in more than a decade whom Germans *have* heard of and whom the caricature resembles as much as, if not more than, it does Tina Yothers–Charlene Tilton of *Dallas*. Okay, I'll stop.

guages not only distinguish kinds of possession, but do so in ways less tidy than a simple alienable/inalienable distinction. The specifics vary from language to language; however, in Maori, there is a class corresponding partly to "inalienable" possessions but, instead of containing body parts and relatives, it contains relatives, food, and small portable objects.

The "sense" one can make of this is that the group contains the things one has control over. This core signification of "control" is alive in Maori speakers' "sense" of the language, and to really speak it is to know when an object is allowed to "visit" the "wrong" class because of real-world gradations in "control." The "controlled" possessive marker is *a*; the "other" group is *o*. *Pou's medicine* under normal circumstances is

te rongoa **o** Pou
the medicine of Pou

because medicine is not something one "controls" in the way that one "has the heart of" an uncle or, apparently in the Maori mindset, a chicken leg. But if Pou *made* the medicine, then Pou is indeed "in charge" of it, and you say:

te rongoa **a** Pou
the medicine of Pou

Okay–but the classification is extremely messy around the edges. Body parts are in the "uncontrolled" class, as well as clothes, grooming implements, and vehicles of transportation! "Control," then, only takes us so far here. Culture did play some role, albeit historically, in what today looks arbitrary: in traditional Maori society, canoes were communally owned and thus not individually "controlled." But then this led to new instances where you "just have to know"–new vehicles introduced by Westerners tend to be classified as "uncontrolled" by analogy with canoes, even though a Honda tends to be individually "controlled." Or–horses, as "vehicles," are usually marked with *o* as "other," but a speaker was once heard to refer to someone's Clydesdales as marked with *a*, because these are horses explicitly trained for show and formations, thus "controlled" in a way obvious enough to motivate a "bending" of the *o*/*a* rule.

All Languages Are Not Equally Complex: What Makes the Difference?

Some languages are by chance more "accreted" than others. It is sometimes said that all languages are equally complex, but more properly, *all languages are complex to some degree.* Overall, a language like Fula is much *more* "evolved" into uselessly baroque elaborations (the verbs are as maddeningly complex as the nouns we have seen) than Maori or, more to the point, English. The Native American language Cree is packed to the gills with "extras" to the extent that an English speaker wonders how a child could even learn it. For example, a marker is used on a sentence's verb just in the case where a noun acting on another one would normally be thought of as subordinated to the other but in this instance happens not to be. This bears illustration:

> Mac- a:yi:siyiniwe:sah nipah-**ik** o:hi ihkwah.
> bad- person kill this louse
>
> "This louse killed the bad person."

The function of *-ik* is only to show that, despite what you might think, the louse killed the man and not vice versa. Or–when mentioning an action, one must use it with a suffix specifying just how the action was done. One does not simply say that a person "chops" but that he chops it with a sticklike object (*kîsk-**ah**-am*), with an ax (*kîsk-**atah**-am*), with a saw (*kîsk-**ipot**-am*), or various other ways–there is no simple "kîskam" alone. Indeed, it has been observed that children growing up with languages like Cree and its relatives cannot be said to have full competence in even ordinary language until the age of ten, by which time English-speaking children are engaged in merely adding finishing touches, mainly in expanding vocabulary, to a language essentially nailed down years before.

Language as an In-Joke: "Simple" People, Complex Languages

Despite the guilty intuition one might have that "primitive" hunter-gatherers would speak "simpler" languages, in fact, it is among such groups that languages are likely to be the most elaborated. The linguist expects that a language spoken by Native Americans, Papua New Guineans, or rural Africans will be jangling with bells and whistles one would never expect from forming one's sense of what a "Berlitz" language is about.

We think it's bad dealing with genders for German nouns–but it's in the Luo language of Kenya that millennia of sound changes have left a situation in which each noun's plural marker must essentially be learned by rote, with only very broad rules of thumb applying. Try to get any sense of how to pluralize a noun in Luo with a list like this:[9]

SINGULAR	PLURAL	
mac	*mec*	fire
gweno	*guen*	fowl
wendo	*welo*	visitor
cula	*culni*	island
saho	*sehni*	skin bag
bawo	*bape*	plank
tigo	*tike*	bead
pino	*pinde*	wasp
or	*oce*	brother-in-law
dak	*degi*	pot
dicwo	*cwo*	man
jamer	*jomer*	drunkard

There are a few places to grab onto: there are four main ways to form the plural, with one (the *-e* ending) more common than all the others. Some of the consonant changes are actually counterintuitive yet regular alternations of the sort we saw in Fula: lots of plurals ending in *-e*, for instance, change *w* to *p* as in *bawo/bape*–a Luo speaker "feels" this as natural in the same way that we "feel" that the final *-f* in *leaf* becomes a *v* when pluralized as *leaves*. But then there are plenty of exceptions, subclasses, and caveats to the "basic" way one forms the plural in all four groups, words where

9. Linguists and Luo speakers: I have omitted indications of vowel quality and length, because they are of no import to the demonstration and would distract from the point of the illustration.

the expected consonant mutations don't apply for no apparent reason, and many words pluralizing in ways other than "the big four" (the last two on the list are examples). Finally, the "big four" ways do not correspond to "gender" classifications as do the Swahili and Fula genders (albeit roughly even there); you just have to know what plural strategy a given noun uses.

Languages that have submitted to a great deal of second-language learning, something inherent to a language's becoming geopolitically dominant, are hindered from developing as much such frippery by the slight pidginization that adult learning enforces. Thus it is predictable that Celtic languages like Welsh, traditionally largely spoken by natives and learned only by the very occasional outsider, have Fula-like consonant mutations that appear to come out of nowhere. In a "well-behaved" Welsh, once you knew that the word for "cat" was *cath* and then you learned that the word for "their" was *eu*, "my" was *fy*, and *ei* was either "his" or "her," then you would just have:

eu cath "their cat"
fy cath "my cat"
ei cath "his or her cat"

But no, this would in reality be "pidgin" Welsh. Actually, the initial consonant of *cath* changes, depending on what comes before:

eu cath "their cat"
fy ***ngh****ath* "my cat"
ei ***g****ath* "his cat"
ei ***ch****ath* "her cat"

Notice that *his* and *her* are distinguished not by the pronoun but solely by different substitutions for the initial *c-* in *cath*–as if in English we said *his cat* and then, to say "her cat," we said *his khat*. Each consonant has its own array of changes like this that particular pronouns exert:

eu plant "their children"
fy ***mh****lant* "my children"
ei ***b****lant* "his children"
ei ***ph****lant* "her children"

Or–in Welsh the verb comes first, such that to say *The dog saw* is:

> Gwelodd ci.
> saw dog
> "The dog saw."

To say *Alun saw a dog*, however, one must transform *ci* into *gi*:

> Gwelodd Alun **gi**.
> saw Alun dog
> "Alun saw a dog."

The consonant change here marks that the dog is now an object. But given context, even without the change, who would have supposed otherwise? This is just a persnickety requirement that Welsh happened to develop and would most likely have either never arisen or been long ago lost if the Welsh had happened to come to hold political sway over vast swaths of Europe and forced their language on adults not used to distinguishing between *his* and *her* by changing the initial consonant of the thing possessed, rather than simply having separate words for *his* and *her*. (How do they mark newlyweds' towels?)

Luo and Welsh are reminiscent of encountering a clique of undergraduates who have been hanging together for a year or more or an office staff that has had relatively low turnover for several years. Groups like these often speak in in-jokes that have acquired an almost impressive degree of richness and subtlety, based on long periods of common acquaintances, trips away, joys, tragedies, chance occurrences, observations–a shared library of experiences and verdicts on them. (The characters on *Friends* display a version of this kind of interaction–all Ross has to do is mumble "We were on a break" and all the characters instantly respond on the basis of the plotline within which this utterance was so poignantly embedded, the studio audience screaming along as they, too, are in on "the joke," having spent so many years with these people.) Only with intimates can one interact on this level, where one has the benefit of such a volume of lore "on-line" at all times. One's interactions with those outside this group will necessarily be based on a less detailed common base of knowledge and references.

In the same way, a language is likely to develop a highly subtle way of distinguishing possessions, an imposing multiplicity of consonant-mutation rules, or a large number of weird sounds most foreigners have trouble mastering as adults, when it is spoken by a group of people whose interactions are largely restricted to other speakers of the language, especially in relatively small communities. In light of this, Fula, spoken over a large area, varies considerably in the specifics of its complexities from place to place.

This means that, if sociohistory thrusts a previously isolated language into a more interethnically dominant role, most likely the language will be simplified a tad over the generations. We can even see the transition from "How do they even speak this?" to just Berlitz-language-style "Dammit!" in previously "indigenous" languages now undergoing heavy second-language acquisition in multiethnic cities. In Senegal, Wolof, formerly one of many languages spoken by groups in the region, has become the urban lingua franca. Related to Fula, Wolof has a similar leaning toward arbitrary classification of nouns into a wide array of genders. Today, in Senegalese cities such as Dakar and St. Louis, there are as many people whose original languages are Serer, Mandinka, Soninke, our friend Fula, and other languages speaking Wolof every day as there are people who learned it as a first language. Where English-language newspapers are full of complaints about people saying things such as *between you and I*, the complaint one hears most among native Wolof speakers in Dakar is that young people tend to just use the article *bi* with all nouns, instead of bothering to choose the proper one out of the many. As Wolof steps out into the wider world, some of its most rococo features are being cleared away. In the same fashion, because most Welsh speakers now use English as their main language, some of the funky consonant changes are used less and less.

People who only speak "kitchen" Maori or have learned it as a second language as part of the movement to rescue the language from extinction at the hands of English are often unsure about whether to use *o* or *a* with a possessed item. The younger generation of Luo speakers, having adopted Kenya's official languages Swahili and English, are losing some of the more arbitrary aspects of the grammar such as the only vaguely predictable plurals, ironing such things out into more regular patterns. Having been adopted as a second language by millions of adults, Swahili is the "easiest" Bantu language, and actually, as Algonquian languages go, Cree is "easy" in regard to variety of endings, most likely because its original speakers migrated into a territory occupied by others, leading the conquereds' nonnative Cree to become an essential part of the mix that children born into the new community heard.[10]

Thus left to its own devices, a human language will tend to elaborate into overt expression of subdivisions of semantic space that would not even occur to many humans as requiring attention in speech and become riddled with exceptions and rules of thumb and things only learnable by rote. This process tends to achieve its most extreme expression among groups long isolated, but any language that has been spoken for tens of thousands of years exhibits some considerable degree of "developmental overkill." It is this feature of human language that contributes to why learning other languages as an adult is such a challenge. No language has been goodly enough to remain completely tidy and predictable, no language has not stuck its nose somewhere where it didn't really need to go, no language classifies objects and concepts according to principles so universally intuitive that any human could pick them up in an afternoon, and in none of them are these classifications indexed to currently perceptible cultural concepts in anything better than a highly approximate manner.

The Only Languages (Virtually) Without Barnacles

Throughout this chapter I have said that any language that has existed for millennia cannot help having accreted layers of decorative "gunk." What this means, though, is that there do exist some languages that are unique in displaying relatively little of this sort of bric-a-brac–namely, the only languages that have not existed for millennia, because they "started again" only a century or five ago: creoles.

---

10. This is a hypothesis of my colleague Richard Rhodes (no, not the other Richard Rhodes).

Pidgin History: Clearing Out the Stables Creoles are usually "born again" from pidgins. Recall that pidgins develop when adults needing to learn and use a language quickly for passing, utilitarian purposes strip it down to its bare bones, eschewing almost everything that is not strictly necessary to communicating. This process naturally means that most or all inflections have to go–after all, if the Chinese haven't needed them in millennia of conversations among billions of people in a real language, then you certainly don't need them to say things like *It's there* and *Ten cents, please.* Tones, as something particularly difficult for adult learners to pick up and not strictly necessary to a viable way of communicating, are out the window in a pidgin, too. In the mines of South Africa, men speaking various languages developed a pidgin Zulu, and though Zulu is dripping with tones, the pidgin, Fanakalo, has nary a one. And then there are the myriad types of linguistic gewgaw we have seen in languages throughout the world that, like earrings, are cute but functionless; predictably, pidgins tend strongly to toss them right out. There are no pidgins with alienable possessive marking, inherent reflexivity marking, consonant mutation, and the like. Not only when the language being learned has such things, but even when the learners' native languages have them, the guiding purpose of maximum learnability and kitchen-sink communication ensures that none of this makes it into the pidgin.

Then the pidgin is transformed into a creole, its material expanded into a real language. The creole certainly is based on a systematic grammar and contains nuanced uses of words, as we saw with Tok Pisin and other creoles. Both of those things are necessary to human communication. However, because the pidgin had so few needless structural "doodads"–and because such things emerge only through centuries and often millennia of gradual "morphing" by sound erosions and changes, grammaticalization, rebracketings, and semantic changes–a creole, having existed for five hundred years max, has yet to amass much "crud." The real language that emerges from a pidgin, then, is not one like Italian, Fula, or Maori. Creoles are, instead, the world's only languages that combine having little or no inflection, little or no tone distinguishing words or expressing grammar, and only a modest amount of complexifications exceeding what one needs to express oneself as a human being.[11] In other words, creoles are the only languages that present the learner with few "Dammit!" moments.

In fact, there is a diagnostic by which a Martian could identify a language as a creole–one that had "started again" from a pidgin a few centuries ago–despite knowing nothing of the history of the language. This diagnostic cannot simply be "languages with neither inflections nor tone that distinguishes words or expresses grammar," because there are age-old languages with this profile, such as the Polynesian ones and some others in Southeast Asia (Cambodian is one), as well as some in West Africa (such as Mandinka and its close relatives Dyula and Bambara), that come close. Yet neither would it be suitable to refine this by adding ". . . or 'Dammit!' constructions," because linguists to date have neither compiled a comprehensive list of all of the possible permutations of needless complexity that a grammar may drift into nor made any unitary characterization of all of them that would also apply to any later discovered.

Yet there is one "crud" feature that just about all older languages appear to have developed by the inexorability of, specifically, semantic change. Often in a language, the meaning of combinations of prefixes or suffixes with roots is so far from what the actual components would mean by themselves that the exact pathway of implications behind the current meaning is unclear, the only sure thing being that a new word has been created from what began as one word and an add-on.

A little scene from a German translation of another one of the adventures of the little Gaulish warrior Asterix illustrates the heart of the matter nicely (see page 208). In this sequence, Roman soldiers charged with vanquishing the Gauls have been sent an all-female relief squadron, much to their amusement. The translator

---

11. Tones can both distinguish words and express aspects of grammar. For example, in the Nigerian language Edo, *ìmà* means "I show," *ímà* means "I'm showing," and *ìmá* means "I showed." A language can also use tones along with inflections, such that the tone is not carrying the load of distinguishing meaning by itself. Some creoles use tone in this way; for example, in Saramaccan, high tone generally falls on the stressed syllable: "house" is *wósu*, where the first syllable carries not only the stress but also a high tone.

has fun in using a series of words combining the root *Lösung,* having the basic meaning "loosening" in these combinations, with different prefixes. I have put in bold the words we will look at here and their English equivalent in the translations:

What kind of masquerade is this?

This masquerade is **relief** for you pathetic dwarves.

Ha-ha-ha! That might be a **redemption** for the troops' morale!

Hee-hee-hee! Stop, or this will lead to my **falling to pieces!**

Ablösung combines the root with the prefix *ab-*. The basic meaning of *ab-* is "away": *abgehen* "to go away." But it is hard to see how *Ablösung*, the parts of which appear to signify "loosening away," comes to mean "relief" in the sense of supplementary assistance. This meaning arose by a series of gradual reinterpretations, just as *silly* came to mean "foolish."

Now look what happens when we combine *Lösung* with the prefix *er-*. With verbs, often this prefix conveys a sense of achievement: *arbeiten* "to work," *erarbeiten* "to gain (something) by working at it." However, *Erlösung* means "redemption," not "to gain by loosening," a meaning it is hard to imagine any culture having much use for. If you really work at it, you can conceive of how redemption could constitute a kind of "heightened loosening"–but redemption is only one of many senses that "heightened loosening" could have come to refer to–"release," "exhalation," or "unraveling" would have been just as plausible. The particular choice that German made was a brand of narrowing of meaning that one could not predict from the initial combination of *er-* and *Lösung*.

Then there is *Auflösung* "falling to pieces": *auf-* is the closest German equivalent to our use of *up* with verbs, even paralleling our *up* in the sense of *eat up*; that is, eat to completion (*aufessen*). As such, *Auflösung* makes some sense as a "complete loosening"–but then German does have a prefix that specifically conveys "falling or breaking to pieces or into ruin": *zer-* (*zerfallen* "to fall into decay"). But there is no *Zerlösung*–instead, *Auflösung* oozed into that meaning even though *auf-* does not literally convey that sense of crumbling. Or–couldn't *Ablösung* and *Erlösung* have been just as plausible candidates for meaning "falling apart" (that is, "heightened loosening")?

None of these questions mean anything to a modern German speaker, who just uses the words "on-line" as we use our own without worrying about the fit between the meanings of their parts and the meaning of the word as a whole. But ultimately, these meanings came from idiosyncratic meanderings through millennia. In an age-old language, conventionalizations of this kind create innumerable cases in which the combination of a prefix or suffix and a root are unpredictable. On the subject of English's good old *understand*, for instance–what's being stood under? Thus though some languages have evidentials, whereas others don't, and some languages have alienable possessive marking or the like, whereas others don't, it is essentially as inevitable as the fact that iron will rust that in time a language will have developed semantically "messy" prefix/suffix + root combinations.[12]

In reference to creoles and whether they can be defined structurally, we even see cases like *Ablösung* and *understand* in languages without inflection or tone as markers of their having existed for more than 100,000 years. In Maori, *whaka-* is the "makes it change" prefix, as in *ako* "learn," ***whaka****ako* "teach." But then there also are cases where you "just have to know," such as *uru* "enter" but ***whaka****uru* "assist" or *tuturi* "kneel" but ***whaka****tuturi* "be stubborn." In a language called Chrau, in Vietnam, the equivalent prefix is *ta-*. No problem with *chuq* "to wear" versus ***ta****-chuq* "to dress someone" (that is, make them wear something), but go figure when it comes

---

12. I know of only one language, Soninke of West Africa, where the grammars and dictionaries do not indicate cases like this, although it has inflections and some tone that distinguishes words and encoding grammar, and thus reveals itself as an old language rather than a creole. But then none of its grammars happen to concern themselves with this kind of "messiness," and none of its dictionaries are very substantial. It may well be that the language has its *understands* just like other languages.

to *păng* "to close" versus ***ta***-*păng* "to close something by accident" or *chĕq* "to set down" and ***ta***-*chĕq* "to slam down," where the combinations have moseyed into highly particular meanings one could not predict from the prefix and the verb.

In pidgins, however, these prefixes (or suffixes) do not "come through" any more than inflections and tones do. Unable to, or not needing to, learn the actual language, pidgin creators tend to strip a language down to its bare roots or, if they recruit a word composed of a prefix or suffix plus a root, it is in "undigested," fossilized form–Tok Pisin has, for example, *insait* for *inside.* However, there is no independent word *in* in the language, with *long* (from *along*) having taken the place of most prepositions; nor is there a series of other words using *in-* as a prefix with a consistent meaning along the lines of *inject, inhale,* etc. The Tok Pisin speaker does not "feel" *in-* as a separate unit of meaning from *-sait*–in Tok Pisin *insait* is one indivisible word (unless one happens to have learned English, in which case one may then see *insait* in a new light).

Thus, in many cases, a creole must develop its own strategies for lending new shadings to a root. Naturally, when these strategies begin, they "make sense." For instance, Tok Pisin has recruited *-pasin,* from *fashion,* to create abstract nouns. Because Tok Pisin has only existed in any form for about 175 years and as a creole for little more than a century, the meanings of these abstract nouns are nicely predictable from the roots:

gut	"good"	*gutpasin*	"virtue"
isi	"slow"	*isipasin*	"slowness"
prout	"proud"	*proutpasin*	"pride"
pait	"fight"	*paitpasin*	"warfare"

There would be no motivation for someone to consciously concoct a "messy" *-pasin* word like, to make up an example, *gutpasin* meaning something like *husbandry*; if someone chose to for some reason, no one else would understand it and it wouldn't catch on. Things like *Erlösung* and *ta-păng* only emerge slowly through time, by step-by-step meanderings of the meanings, each little step making transparent sense to a living speaker of the moment. For example, in the literal sense, *innumerable* "should" mean "unable to be counted," but it hardly throws us that the word is actually used in a slight extension of that meaning–"of great number," because this is a probable concomitant of not being able to be counted. However, in regard to *understand*, in our lives we have encountered the word after so many incremental movements away from the "stand under" meaning that we cannot even relate the modern meaning of *understand* to what its parts "should" mean.

In the same way, check up on Tok Pisin in a thousand years, and *isipasin* may have morphed slowly into meaning perhaps "stupidity," and so on. But that kind of thing takes eons, and we catch the language at an early stage of its history when semantic drift of this kind has yet to occur to such degrees.

In many creole-speaking societies, people have often also spoken the language that provided the creole's words, especially after the abolition of the slave trade. Thus Jamaican "patois" speakers often speak English as well; a small but influential minority of Haitians have always spoken French along with creole. In cases of this kind, the inevitability of language contact ensures that the creole eventually takes on some of these "messy" affix-root combinations. For example, in French there are many such uses of *re-*, such as *rejeter*, where *jeter* means "throw" but *rejeter* means "to reject," despite rejection not necessarily entailing one's throwing something away *again*. French creoles like Haitian, Martiniquan, and the one in Louisiana have always been spoken by many people who also speak French, and thus these creoles have *rejete* "to reject" and its ilk, having inherited words from French just as English did after the Norman invasion.

But there are many creoles where history has it that their speakers did not speak the languages that provided their words, such that the creoles have developed largely from their own materials. Tok Pisin speakers rarely spoke English until recently with the migration of many people from the bush into cities. The creoles of Surinam, such as Sranan and Saramaccan, developed from a base of English words when the English colonized Surinam from 1651 to 1667; after this, however, the Dutch ruled the colony until 1975, and thus few speakers of these creoles had any significant contact with English for centuries.

Thus creoles like this are the ones that a Martian could identify as having been born recently from pidgins: any language in the world that has no inflections, tone to distinguish words or express grammatical features, or semantically unpredictable prefix/suffix + root combinations is a creole.

Creoles, Children, and Sign Languages: An Instructive Intersection One indication that creoles represent human language closer to its core essence than old languages do is what results in the very occasional situation where children, who have not already been exposed to an old language full of "dings" and embellishments, transform a pidgin into a real language in one generation. What they create is a creole, rather than a language like Cree or Vietnamese. Usually, it is adults who gradually expand the "just for now" scraps of a pidgin into a fuller vehicle with the rules and flexibility we associate with a real language, something that can be spoken "wrong" as well as right. Tok Pisin, for instance, existed long before many children were using it as a primary language in a form about ninety percent as "gelled" as it is today among native speakers. But every once in a while, social history has created the bizarre circumstance in which children took a pidgin and transformed it into true language.

In turn-of-the-twentieth-century Hawaii, for instance, Portuguese, Chinese, Japanese, Filipino, Korean, and Puerto Rican immigrants were speaking a useful but rudimentary pidgin English, varying in its structure from speaker to speaker, depending on a person's native language and its effects on how that person rendered the new language. *Gud, dis wan,* an Ilocano speaker from the Philippines would say–"good, this one"–because his language puts the subject at the end of a sentence. But the Japanese person, whose native language puts verbs at the end of the sentence, was saying things like *Mi kape bai* "my coffee buy" for *He bought my coffee.*

The immigrants' children, while learning their various home languages, played together using mostly English words as the common coin. Because they were children, however, the human predisposition to have a full language played its hand, and they transformed the just-getting-by pidgin into a real language, with everyone using the same rules regardless of the language they learned from their parents at home. *Who wen cockaroach da orange juice from da icebox aftah I wen kapu am?* a speaker of the local creole (misleadingly called "Pidgin") might say today, where *cockaroach* is "steal," *kapu* is "to make forbidden," and *wen* is a past marker (originally *been*), quite foreign to Standard English. Importantly, a Hawaiian Creole English speaker would use these words and constructions regardless of their specific ancestry. Aspects of the sentence show that this is a creole English–the preverbal marker to indicate tense instead of a suffix, as well as the collapsing of *him, her,* and *it* into the pan-gender *am*. All speakers of the creole would also say *Dey stay run* for *They are running, stay* having becoming a preverbal marker of the progressive analogous to English's *-ing* suffix.

If languages like Luo were "the heart of language," then what the children in Hawaii created would presumably have had gender markers, conjugation, maybe some tones, and God knows what else along the lines of alienable possessive marking, unpredictable plural marking, etc. But instead, Hawaiian Creole English fits the broad structural profile of languages like Tok Pisin, Sranan, and Angolar–what the immigrants' children created in Hawaii is a language that "started again": a creole.

The Hawaii scenario is a far from perfect "laboratory case" of language starting anew in one generation, however. First of all, the children had learned their parents' native languages from birth at home and, at the same time, were going to school, where they were being taught, and taught in, Standard English every day. Also as a result of this, there is more English in the mix in Hawaiian Creole English than in Sranan or Tok Pisin; it falls somewhere in the middle of the continuum of creolization we saw exemplified by Guyanese. Surely the fact that the children did create a creole of any kind as their in-group lingua franca is fascinating in itself, in showing the power of group / peer identity over formal tutelage when it comes to language. Nevertheless, it cannot be said that these children truly created a language by using only the pidgin materials as a source, nor is there any context past or present where the evidence suggests that this ever happened. Such a situation would require the rather eldritch circumstance of parents not passing on their native languages to their children, instead only exposing them to a language they spoke only haltingly and were not comfortable in.

A true demonstration case of language as created from the ground up with no interfering input from preexisting languages comes from the only conceivable situation in which groups of children could grow up in healthy circumstances while somehow not learning their parents' native language. If you think about it, the only possible way this could happen would be among deaf children, and indeed just such a scenario has been carefully documented for the past twenty years in Nicaragua, where in the 1980s deaf children created a new sign language while attending a new school for the deaf. Before the school was founded, there was no established sign language in Nicaragua and deaf people had little contact with one another. Each of the children had been using a "home" sign language local to their households before they came to the school–a sign language largely based on the kind of manual mimcry that most of us would come up with if forced to communicate with our hands only. Once the children were brought together into the school, however, they quickly conventionalized a systematic sign language of their own capable of expressing all human thoughts.

Sign languages are "real" languages just like spoken ones, with grammar, complexity, and nuance; second-language speakers even contrast markedly with native ones in regard to fluency, thus having an "accent" in their sign language. I visited the school where Nicaraguan Sign Language arose and had ample opportunity to see how very much of "a language" it is, right down to watching teens flirt with and tease one another in it. I was given a "name," a finger placed over the eyebrow, referring to my thick eyebrows. I will never forget the sight of a whole room full of children animatedly communicating all of the things that any group of kids would to one another–in perfect silence broken only by the occasional swish of fabric that moving one's arms occasions.

Yet in its structure, Nicaraguan Sign Language in many ways parallels neither Spanish nor languages like Welsh and Luo, but creole languages–for the very reason that it is a new language just as creoles are. In fact, because all of the world's sign languages are of relatively recent origin–for example, American Sign Language was developed about the same time as Tok Pisin, in the 1800s, and resembles that language in structure much more than it does English–these languages can be argued to be "manual creoles."[13]

But Language Hits the Ground Running Most of the world's languages are so full of "Dammits," then, because use through millennia inevitably submits them to varieties of language change and inexorable yet useless "baroqueification." Even creoles, because

---

13. The parallel is not exact, however. For example, the exigencies of manual versus spoken expression lead sign languages to make use of noun classifiers along the lines of Cantonese and other Chinese languages, despite the fact that among spoken languages these are a minority phenomenon, surpassing what is strictly necessary to vocally based communication and arising through random elaboration.

they have existed for longer than a year or two, exhibit certain degrees of such elaboration.

For one, regularly giving overt indication of tense is technically "overdoing it" as far as communicative necessity is concerned. Many old languages indicate tense only when absolutely necessary, such as on the first verb in an oral "paragraph" and then leaving subsequent verbs bare, or they even never indicate tense at all when context does the job. (Chinese speakers will find this familiar.) A few languages even go as far as not to have any way of regularly indicating tense–no *-ed*, *-te*, or *-ó*; instead, if you really must pin down the time, then words like *yesterday* and *tomorrow* are used. This is found, for instance, in some languages of Australia and Papua New Guinea.

Creoles, however, always have their suites of preverbal particles to indicate time of occurrence, in many cases using them in the first way just indicated, marking the time at the beginning of the oral paragraph and then often leaving it off after that. Even creoles, then, technically gild the lily a bit in regard to how much semantic space they fill in. This shows that even creole creators did not restrict their new languages to the *very* core of what a language strictly needs to contain. Certain elaborations of the source languages had a way of nudging creoles at least a small distance from the true sine qua non, even if this did not go as far as including gender markers and evidential markers. The same point could be made with definite and indefinite articles, which creoles often have both of.

Thus, left to its own devices, a language will develop baubles–linguistic overgrowth that, whatever its interest, is incidental to the needs of human exchange and expression. Heavy amounts of learning by adults have a way of retarding some of this faceless overdevelopment in the interests of general user-friendliness.

But under other circumstances, the whole of a language's machinery is retarded in its natural bent toward eternal transformation. One need not travel to New Guinea or Surinam to witness this phenomenon. In fact, your native language is most likely just one of these languages that contingencies of social history have rendered "evolutionarily challenged."

6

Some Languages Get Genetically Altered and Frozen

I have emphasized that spoken language is an ever-changing system, the very nature of which is to be always in a process of transformation into a new language. This can be difficult to perceive because the process is such a gradual one–to make the point for English, for example, we must look back to *Beowulf* and *The Canterbury Tales* and wrap our heads around the fact that the language we speak did not yet exist when they were composed. Yet in some societies the inherent impermanence of language is much more obvious within the span of a human lifetime than in Western ones.

Language Is a Lava Lamp

Language Change as It Is Really Lived

In the northern Australian language Ngan'gityemerri, in 1930 the way one said *He poked along, tracking it along here to where it made its camp* was:

Dudu	dam,	dam	dudu,	kinji	dinj	parl.
Track	poke	poke	track	here	he-sat	camp

To express "he poked along, tracking it," one juxtaposed the verbs for *track* and *poke*; in this sentence the speaker happened to use the verbs in both orders, *dudu dam* and *dam dudu*.

Yet today, the way in which a Ngan'gityemerri speaker (actually, the Ngan'gimerri dialect of Ngan'gityemerri–"Dialects Are All There Is") would render this thought is:

Damdudu, damdudu, kinyi dinyparl.
Poke-track poke-track here he-sat-camp

The difference is that today, instead of juxtaposing *track* and *poke* in whatever order one chooses, one must use a single verb complex "poke-track." No longer can one say "track poke"–today, *poke* has become a prefix added to *track*, such that *damdudu* means roughly "He pokingly-tracked." One can no more say "track-poke"–*dudu dam*–than an English speaker could say *paddle-doggie* instead of *doggie-paddle.*

Furthermore, the poke-tracking is not an isolated funny little expression but an example of a general pattern that has gelled and set in the language in just the past seventy years. Today, the Ngan'gityemerri speaker must express how one did *anything* as a prefix rather than as a separate verb placeable either before or after the main verb itself. Thus from 1930 to today, the language has gone from being one with a predilection for stringing verbs together like many African languages (that is, why Sranan creole, modeled on African languages, has *That hunter* ***bought*** *a house* ***gave*** *his friend* instead of *That hunter bought a house for his friend*) to one with a predilection for cramming into *single* words with several "pieces" what English speakers expect two or three words to cover, more typical of Native American languages like Cree (for example, Cree's *kîsk-**atah**-am* "he chopped it **with an axe**"). In the same way, whereas the Ngan'gityemerri speaker in 1930 would have said simply *He sat camp,* today there is a single word *sit-camp,* such that today's *dinyparl* translates roughly as *He camp-sat,* like our *babysat.*

Moreover, in Ngan'gityemerri one is virtually *obliged* to use some kind of prefix of this sort to express just how one does almost anything, just as in Cree one must specify how one chopped something by adding a suffix to the verb. In the old days the Ngan'gityemerri speaker expressed how one did something in this way *if it felt necessary,* as is true with an English speaker–we say "Who's going to chop the carrots?" not "Who's going to chop-with-a-knife the carrots?" However, the modern Ngan'gityemerri speaker *must* specify "all the way" whether he feels like it or not–

the language has oozed into obligatorily filling semantic space that many languages have little trouble leaving unfilled most of the time.

To young Ngan'gityemerri speakers, the old way of speaking is something they remember of old people now departed–that is, the people who were recorded in 1930–but they would never speak that way themselves. In other words, not just the slang but the basic structure of the young people's language is radically different from that of their grandparents and great-grandparents. The linguist investigating this language in the 1990s was no more encountering the language as it was spoken in the 1930s than an interviewer speaking to George Burns in the 1990s, a hunched-over little senior croaking out one-liners about being old, was encountering the George Burns of the 1930s, strutting around erect and unlined in black and white on vaudeville playing straight man to his wife, Gracie Allen.

Language Change for the Man in the Gray Flannel Suit

To an English speaker, language change this rapid is unthinkable. Sure, English changes, but for us, the English of five hundred years ago, such as that of Shakespeare, is quite recognizable as the language we still speak. The French speaker needs to make only relatively minor adjustments to read the sixteenth-century essayist Montaigne; and Martin Luther's 1522 translation of the New Testament into German is accessible to today's Germans.

Yet time was that even the Western languages changed faster than they do now. Although for us Shakespeare's language is taxing but hardly Hebrew, Shakespeare would have had to study the English spoken five hundred years before he lived as a separate language. In A.D. 1000 English had been an inflection-laden tongue so unlike what we think of as English that, by the late 1400s, printer William Caxton, frustrated by the variegation among English dialects and briefly considering referring to Old English in search of a printing norm, rejected it as "more like to German than English."[1] To us, Shakespeare would have talked really funny; but to

---

1. What he actually wrote was "more lyke to dutche than englysshe," which almost looks more like German than English to us now! (*Dutche* meant *German* in the English of Caxton's time–*Deutsch* is *German* in German.)

Shakespeare, an Old English speaker would have been utterly incomprehensible. Since A.D. 1500, English has changed, but nowhere near on the order of housewifes becoming hussies, *God Be with You* becoming *Goodbye*, or *ever each* becoming *every*. Similarly, Martin Luther, whose Bible is parsable to modern Germans, would have found the German of A.D. 1000 all but opaque.

Why did languages change so much faster in earlier eras? In fact, a more appropriate question would be: Why do languages change so much more slowly nowadays? The fact actually is that a small subset of (mostly) Western languages today change much more slowly than languages are "supposed to." From *Beowulf* to *King Lear* is *normal* in relation to how fast languages tend to change throughout the world. The Ngan'gityemerri example is an extreme, but still. Polynesian people began their dispersal across several islands (or island complexes) such as Tahiti, New Zealand, and Hawaii about 500 B.C.; it did not take much longer than five hundred years before the original group's language had evolved into new, mutually unintelligible ones in each of the new societies their descendants had founded.

Or–just two hundred years ago in the late 1700s, when missionaries recorded the Saramaccan creole spoken in rain forest communities by descendants of slaves escaped from plantations in Surinam, the language still had consonants between vowels that today are gone, the two vowels now simply juxtaposed. Today *woman* is *mujéɛ*, where the two *e*'s signify two syllables, high tone occurring on only the first *e*. In the late 1700s the word was more like *mojéri*, where an *r* still intervened between the two vowels (this word is derived from Portuguese's *mulher*). To negate a sentence in early Saramaccan, one used *nó*, but today it has evolved into *á*, something that an eighteenth-century Saramaccan speaker would find otherwordly.

In contrast, as I write this I am reading a lovely coffee-table book on American presidents called *To the Best of My Ability*. All of the inaugural addresses are provided in the back of the book, stretching across the exact same period of time during which Saramaccan has changed so much. I am struck by the fact that George Washington's first address of 1789 is in very much the same language as Bill Clinton's second one of 1997. Listening to Washington deliver his speech would require no significant adjustment of us, even though a Saramaccan child having traveled back in time to listen to a speech given by a chief in 1789 would pester his mother asking why the man talked so funny.[2]

What crucially distinguishes the big Berlitz languages from Maori and Saramaccan in this regard is standardization and, most decisively, the widespread literacy that usually followed standardization. Something quite recent and unprecedented in human history has profoundly affected the life paths of a certain few of the world's six thousand languages. First, nationalism and geopolitical accident have anointed certain dialects of these languages as "standard" (or in other cases motivated the creation of a "standard" by combining several dialects). Then, this standard has been transcribed into writing on paper.

This second step is what holds a stick between the spokes of the spinning bicycle wheel of language change. Written language was originally intended to reproduce spoken language on the page. However, whereas speaking is a largely subconsciously controlled activity (we do not actively parse and construct our sentences as we speak), writing is a slow, conscious, controlled endeavor, of a sort that lends itself to conservatism: the ingraining of habits and in-house customs simply because that's the way it was done before.

An analogy is the tendency for cartoon characters to be bipedal, speaking animals wearing gloves. The first thing that strikes one about European cartoons is that they are less likely to be about cuddly talking animals–they tend to be about humans. And really, this is the more "natural" choice of topic in grand view–if you decide to make an animated cartoon, why, precisely, would focusing on four-foot-tall rabbits and ducks living in apartments be your first idea? Indeed, early animated cartoons were as likely to be about people as animals–Koko the Clown, Colonel Heezaliar, Farmer Alfalfa, the last surrounded by animals that were small, feral looking, and largely mute–like most *real* animals! The change occurred as a response to Mickey Mouse in the late 1920s. Walt

---

2. However, inaugural addresses of the olden days contrast with modern ones in length–back in those days, oral delivery was cherished more than now. (I wrote that on November 7, 2000, the night that the markedly inarticulate George W. Bush was elected president.) William Henry Harrison, the ninth president, went on for so long in 1841 (today covering six-and-a-half pages of tiny print tightly packed), trying to show that a man in his late sixties was hearty enough to serve by doing so in the cold without hat or coat, that he caught pneumonia and died after a month in office!

Disney began his career making the natural choice, cranking out a series about a human, Alice (that is, of Wonderland fame), surrounded by typically bestial little animals. But then Disney happened to decide to do some cartoons about an oddly humanoid mouse (after having a similar rabbit character stolen out from under him). The first Mickey cartoons are scraggly, forgettable little jobs, and "Mickey Mouse" would most likely have made little lasting effect if Disney hadn't hit upon making the third entry in the series the first cartoon with sound, *Steamboat Willie.* Because of the novelty of the sound, this cartoon, and thus Mickey, were a huge hit, and this success immediately sparked a flood of imitations–Flip the Frog, Bosko the Talk-Ink Kid, a Krazy Kat that tossed out the existentialist idiosyncrasy of George Herriman's comic strip and recast the character as a Mickey clone.

From then on, cartoon characters at each studio developed rather like "dialects" of Mickey Mouse. As time went by, each studio's characters developed their own signature flavor, but because the development occurred step by step on the basis of the original Mickey template, American characters ultimately remained rooted in that mold, just as the Romance languages have always remained rooted in a Latinate / Indo-European model. The reflex to render cartoon characters as animals about a yard tall wearing clothes and gloves and festooned with a generic smile resulted, then, from simple blind tradition–eventually, the Mickey model became one's default conception of how to design a cartoon character for the simple reason that this was all anyone had ever known for decades. This included the gloves–who, really, walks around wearing white gloves all the time and, especially, what woodland creatures are given to such behavior? Disney had originally provided Mickey with gloves because it made the hands easier to draw; after that, all of us learned to accept gloves on cartoon animals as casually as we accept that typewriter keyboards begin with QWERTY.

In the same way, the prestige of writing as the vehicle of education and its physical constancy in contrast with the ephemerality of, as the late Anthony Burgess put it, a mouthful of air mean that we tend to conceive of written language as the prototype of "language" itself, as how language "should" be. Thus written language does not change at the Ngan'gityemerri "mouthful of air" pace. Instead, the existence of a language in writing tends to lead its speakers, if most of them are literate, to reconceive spoken language as a kind of pale, sloppy reflection of the "real" language on the page. Changes in the spoken language are regarded as a kind of shaggy entropy, a defacement of something considered set, eternal, the alteration of which constitutes destruction. This is our tacit sense of what "English" is, for example–but in grand view this is a highly contingent affair. Cree and Tuyuca speakers neither mentally "see" their languages on a page nor process what comes out of their mouths as a "version" of something "best" expressed by scratches on paper.

That sense that the written variety of a language is "The One," then, is due not to anything inherently superior about that version, but to the seductive power that anything has by sheer virtue of its having been there first, as long as the means are available to keep its image available in perpetuity. Take Mickey Mouse again, for example. Really, there's nothing terribly compelling about Mickey as a character. What's Mickey "like" other than diligent and nice? When have you ever been moved by Mickey as an *individual* rather than by what was happening around him, such as brooms on the march or a truly interesting entity like Donald Duck yelling at him? If anything, Mickey is a rather bland personage compared with Bugs Bunny or Barney the drunk on *The Simpsons,* and this facelessness was well established even as early as *Steamboat Willie,* where perhaps the most individual thing he ever did was whistle. Mickey Mouse is ultimately just one more cartoon character of no God-given distinction. He was imitated by all the other studios for the mundane reason that he happened to be the main character in the first sound cartoon and, for that reason alone, became a sensation. The dialect and time slice of a language chosen to write in feels so anointed for the exact same reason–it just happened to be the one to get immortalized on the page.

But once a society falls under this kind of impression, change as rapid as that between Old and Modern English or between Proto-Polynesian and Hawaiian becomes impossible. Instead, generation after generation are taught that the "real" language is that variety enshrined on the page and that the changes taking place in their speech are "mistakes" rather than natural developments of the very sort that turned Latin into French or some lost language into Japanese and Korean. This tutelage cannot eliminate language change entirely, but it does put a major brake on the process. Thus standardization and widespread literacy, for all of their obvious advantages, retard language change. These processes greatly affect only the relatively few languages that have undergone this process, those spoken by groups who happened to acquire geopolitical juice over substantial regions in the past millennium–only four percent of the world's languages are spoken in, for example, Europe. Thus from a bird's-eye view, standardization and literacy have left the world with a small collection of languages that, in comparison with all of the others, have been curiously inert in their rate and degree of change in the past few centuries.

Writing Sits Still While Speech Moves On: The Real Deal with French

To be sure, writing does not freeze a language in place. There is an extent to which spoken language simply develops apace, leaving the recidivist written language behind.

One of the underacknowledged pitfalls in learning to speak French for a foreign learner is gradually realizing that the distance between the written language and the way it is actually spoken colloquially even by educated people is vaster than textbooks generally acknowledge. One learns, for example, that the way to express *we* is the pronoun *nous*, with its corresponding ending *-ons*, as in *nous faisons* for *we make*. Yet in reality, *nous* has not been used much in casual French for centuries. Instead, it has been largely replaced by *on*, which began as the impersonal third-person pronoun used as English uses *they–on dit qu'il est malade* "they say (that is, it is said) that he is sick." Textbooks will mention that *on* is "often" used in place of *nous*, such that *nous faisons* is "often" *on fait*, but they rarely make it clear that the actual situation is that one simply *cannot* engage in a casual café conversation in Paris using *nous*, anymore than one can speak casual English without using contractions (imagine: *I will tell you, man, I did not have anything to do with it!*). *We talked with her yesterday* in social, wine-with-dinner French is ***On** a parlé avec elle hier.* I will never forget a French-speaking girlfriend telling me in a restaurant after I had ordered a meal and said *nous*[3] several times

3. No, I had not boorishly ordered for both of us; I forget why I would have had occasion to say *we* so often when asking for sausage.

in the process, "Okay, the first thing you have to let go of is this *nous.*" The spoken language long ago replaced *nous* with *on*; one uses *nous* in formal language, in line with the sense that tony situations occasion using the "real" language. But properly, the genuflection to *nous* at conferences and in speeches manifests a dutiful return to a past stage of the language, rather like donning a top hat, watch on a chain, and lorgnette at a *Titanic* party.[4]

Similarly, one does not walk around in France saying things like *Je ne parle pas* for *I am not speaking*; as far back as the Middle Ages, speakers were already dropping the *ne*, such that colloquial French prefers simply *Je parle pas.* One *must* learn this to keep from sounding like an android or, at best, "cute." Just as cartoon characters created in 1950 had the same gloves and smiling mouths as Mickey did in 1928, written French retains this *ne* despite the fact that one must usually drop it when speaking, on pain of dying alone.

Then there's the *c'est* issue. Textbooks lead one to suppose that saying things like *La France est un grand pays* "France is a great country" is default, but gradually one realizes that in running conversation, if a French person really wanted to call attention to how great France is, she would likely say *La France, c'est un grand pays* "France, it is a great country." Whereas word for word this sounds either clumsy or colloquial in English (*My Mom, she's an accounts receivable over at Walmart*), in French this is the way the language is generally spoken, period. If a Martian came down to earth and penned a French grammar by simply listening to people *of any social class* talking casually all day long, he would likely designate *c'est* the third-person-singular verb form of *to be* rather than just *est.* Even Louis XIV demonstrated this, in saying for *The state is me* not *L'état est moi,* but *L'état, c'est moi*–even though it would be inconceivable to imagine George the Third saying, "The state, that's me" as an official pronouncement. It's no accident that in French-based creoles, created by slaves who heard only the colloquial language, the equivalent to *to be* is often *se* (that is, *c'est* as it is pronounced)–Haitian Creole's *He is a fisherman* is *Li* ***se*** *pešer*–and never a word based on just *est.*

4. I assume that various people had such parties. Did you know that even many of the first-class passengers shared bathrooms with other suites?

This means that whereas in a French newspaper *Working is not the only thing we don't do* would be

Travailler n'est pas la seule chose que nous ne faisons pas,

the way anyone would say this at home would be

Travailler, c'est pas la seule chose qu'on fait pas.

There are in effect two Frenches, then, and the unwritten one is quite a different affair from the one on the page. Written French holds onto an earlier stage in French's evolution, having been codified several centuries ago. Yet the coexistence of these two Frenches, and similar divergences in other standardized languages, have a particular by-product: the reification of "The Language" as a changeless entity to be striven toward rather than a dynamic system to participate in as we approach a dance step or cooking. It is this that ultimately becomes a kind of albatross on the spoken variety's neck and holds it back from its normal rate of transformation.

Mickey Ties Us Down: Literacy and Language Change

How this gap retards language change is that the very existence of the written variety has a way of keeping the spoken language from moving along as quickly as it would otherwise, even if it does advance somewhat as spoken French has. Standard English as we know it, for example, is, properly speaking, an embalmed dialect held back from ambling down paths that speech varieties throughout the world have gaily taken to no general misery or discomfort.

"Good English" or Old English?

We are always told that "two negatives equal a positive" and that therefore a sentence like *He didn't see nothing* is "illogical." This is one of many "rules" that must be hammered into us in school–the real reason being that the "rules" have been imposed on the language from without, rather than arising naturally within them (*Billy and **I** went to the store* is another such rule).

The pox on "double negatives" is surely the most utterly silly of these rules, for the simple reason that "double negatives" are the usual situation in languages throughout the world, all of which surely cannot be branded as dwelling in "illogic." ***Nunca*** *he visto* ***nada*** "never have I seen nothing" a Spaniard would say for *I have never seen anything*. This is a double negative and yet would not be out of place in Cervantes. Learning most other languages entails "unlearning" the discomfort we are taught about double negatives, and in fact English is rather odd in expressing negation by "single negative" sentences such as *He didn't see anything*. If you think about it, isn't it a little weird to express that nothing was seen by specifying that what one did not see was anything and therefore that one saw nothing?

Even Old English allowed double negatives:

Ic ne can noht singan.
I not can nothing sing

"I can't sing anything"[5]

Consequently, by the 1500s, when English had become something we would recognize as what we speak, double negation was legal in all of Old English's dialect descendants. However, in some dialects, including the mixture in London that chance thrust onto center stage as the standard, an alternative to double negation had arisen along the lines of our *He didn't see anything* rather than *He didn't see nothing*, which is the way such a thought could, and still can, be rendered in just about any non-Standard English dialect as well as most of the world's other languages. Importantly, though, single negation was an alternative, not the rule. Even within London, double negation was commonplace in nonstandard speech (think Cockney-speaking Alfred Doolittle in *My Fair Lady*), and moreover was even used innocently in standard speech alongside the new "single negation." Examples even pop up in Shakespeare: the

5. Our *not* arose when the *ne* before the verb gradually wore away, just as the equivalent did in French, while simultaneously the "nothing" meaning of *noht* bleached–*terrible* style–into mere negation. Thus *not* arose just as *pas* did in French, beginning as a reinforcement of a preverbal *ne* and ending up being the whole show. *Noht* in its original meaning hangs on at the margins of the language as *nought*, still an everyday word for *nothing* in some British dialects but for us restricted to the frozen expression *for nought*.

character who says "There's never none of these demure boys come to any proof" is Falstaff [*Henry IV* (II), 4.3.97], not LL Cool J. Double negation in this period was an *emphatic* strategy–two negatives, rather than canceling one another out like integers, meant heightened negativity. Never none of those demure boys whatsoever, dude.

Yet John Cleese and Dick Cavett do not have double negation at their disposal as a way of adding a bit of weight to an observation in Standard English, and the reason for this is less natural evolution than the power of the printed page. Responding to a summons by the House Committee on Un-American Activities in 1952, Lillian Hellman famously declared "I cannot and will not cut my conscience to fit this year's fashions"–but cut its conscience English unfortunately did, in the late 1700s. The "fashions" of the time were founded on an inferiority complex about English's "suitability" as a language of newly international note, and in 1794 a certain Lindley Murray wrote a treatise on English grammar, modeled in large part on a similar one published three decades-plus before by a certain Robert Lowth, which can only be described as having hit the ground running. Murray's book was recruited as the basis for what "good English" was supposed to be, used in classrooms on both sides of the Atlantic for decades to come and sparking innumerable imitators.

When it came to double negation, Lowth, followed by Murray, decided that its prevalence in almost all English varieties and its optionality in the London variety was founded on a pervasive fault of logic endemic to the English population, supposedly neglecting that, as he put it, "Two negatives in English destroy one another, or are equivalent to an affirmative." This notion was based partly on the fact that Latin did not allow double negatives–but modeling English on Latin made no more sense than declaring that cats ought not meow because dogs don't. Alas, double negation was nevertheless taught as improper to write and thus, never seen on the page where the purportedly "real" English was enshrined, became processed as ineluctably "lowbrow," which it remains to this day.

Left to its own devices, Standard English would most likely allow double negation as an emphatic strategy, along the lines that Falstaff used it. "This year's fashions" shunted the dialect away from its natural evolution. Non-Standard English speakers everywhere embrace it–Cockneys, Appalachians, colloquial Singapore English ("Singlish") speakers, black Americans, Brits from the Midlands–but as Standard English gradually makes inroads on nonstandard dialects in English and the United States, diluting their most prominent "nonstandard" features, single negation of the *I don't see anything* variety takes on an increasingly large role in the speech of nonstandard speakers as well. Thus single negation, originally a mere optional ripple in the grammar of some English dialects, spreads throughout the English-speaking population like Sargasso weed, while the French, apparently following the old dictum that "Fifty million Frenchmen can't be wrong," can declare that *Ce qui* ***n'****est* ***pas*** *clair* ***n'****est* ***pas*** *français* ("What isn't clear isn't French") by actually *using* an "illogical" double negative *twice* and can moreover even wallow happily in *triple* negatives (*Je* ***n'****ai* ***jamais*** *vu* ***personne*** "I have never seen anyone").

Make no mistake–double negatives sound every bit as Joe Sweatsock[6] to me as to any English speaker; it would be almost impossible to grow up speaking and reading English feeling otherwise. The die is cast–I grew up seeing almost only single negatives on the page and thus internalizing a sense that "*English* English" was strictly the written form. Only intellectually can I claim to have escaped this; as a human being, my spiritual relationship to double negatives will remain forever tainted by the vagaries of sociological evaluation. My point is simply that this natural feeling is an *arbitrary* imposition tracing back to underinformed pronouncements made more than two hundred years ago by disproportionately influential people. It has nothing to do with "logic," which language grammars worldwide gleefully contradict with abandon. (Why don't we say *Amn't* I? instead of *Aren't* I?; Why are German forks women?)

"You was very rude about my bust yesterday. I'm very sensitive about it," says Mrs. Slocombe on one episode of the British sitcom *Are You Being Served?* Mrs. Slocombe is from the "North country" and occasionally exhibits dialectalisms that belie her assumed air of posh origins.[7] Despite its salt-of-the-earth "feel" to us, *you was* is,

---

6. Appellation source: Sideshow Bob, *The Simpsons* (episode 3F08, "Sideshow Bob's Last Gleaming").

7. At a fancy restaurant, a waiter asks Mrs. Slocombe, "A nice aperitif . . . ?" She responds, "Ah, yes, and they're me own, too!"

really, the way an optimally well ordered English would handle things. Imagine:

SINGULAR		PLURAL	
I	*was*	we	*were*
you	*was*	you	*were*
he or she	*was*	they	*were*

What's wrong with that in the end? The only reason Standard English has ever had *you were* when referring to one person is that, formerly, *you* referred only to two or more people, whereas *thou* was the singular second-person form. Gradually *you* came to be used in the singular as well: first, as a "formal" address analogous to *vous* in French and, finally, in all situations and ousting *thou* completely in spoken English by about 1700.

Once *you* was used in the singular, however, it was a natural development for speakers to use *was* rather than *were* with it, because other singular pronouns took *was*. This followed the general tendency in language change of extension, as we saw with the plural *-s* spreading from one class of English nouns to all nouns. This was indeed what happened in the West Midlands and in northern England in particular. Even in the United States, this usage was not unheard of even among middle-class people as late as the 1800s. Here are excerpts from some letters that a young store clerk wrote to an *amour* of his in New York City in the mid-1830s:

> I knew you **was** not serious when you told me that Thursday night, might be the last one which I should pass with you, for I won't believe that you could tell me that, and not manifest any feeling of regret at all on parting.
>
> Indeed, I know not one word you *did* say, for I was so perfectly astonished in the first place, to see you going home without appearing even to think of me, and then when I met you at the door to find out that you **was** angry with me, I knew not what to make of it. There were many people looking at us, and I knew it.

The chances that these usages were slips of the pen are slight, given that they are found amid prose of an elegance beyond the capabilities of most seventeen-year-olds today, and more than once; one also encounters them in letters by other educated people of the period. Indeed, the writer was the son of a lawyer who had been the head of an elite family in respectable Newburyport near Boston. *You was* rather than *you were* was an option available to a Standard English speaker in Jacksonian America.[8]

Yet the deck was stacked against *you was* even as our swain wrote. Lowth and Murray did not approve of the "misuse" of *was* with *you,* and the influence of their books drummed the conditioned habit of *you were* into all succeeding generations of people concerned with speaking "proper" English. It is virtually certain that, if English dialects were spoken in a rain forest somewhere and had never been written down, *you was* would long have been obligatory in the singular throughout all of the English-speaking villages.

What we process as the soul of respectability in the language of our moment, then, is ultimately merely "this year's fashions," with the fashions of a certain year frozen into place by widespread literacy, rather as if, during the Clinton Administration, Hillary Clinton had been required to dress like Martha Washington in all public appearances just because that's the way Martha dressed. Any given language chooses from an infinite array of possible grammatical configurations, on which notions of respectability are arbitrarily superimposed, meaningless to people speaking the language or even dialect next door. People in the courts of early England quite confident of their higher station than the thane in the streets were holding forth with double negatives galore. Just 170 years ago, a callow lovelorn fellow-about-town seeking the affection of a woman well known for the cut-glass elegance of her letters did not see *you was* as crabbing his message in the least. But today, codification having decisively frozen *you were* in place, the same *you was* renders any written prose either dialectal or juvenile such that, when the book those letters were taken from was in editing, the author was surely asked whether the *you was*'s in these letters were perhaps a mistake, and a suitor, if insisting on using *you was* today, would be required to compensate for it by sheer charisma or other attractants. John F. Kennedy, Jr., could not have e-mailed his Carolyn that *There wasn't nobody that was looking as hot as you was last night,* because the English he spoke had been suspended in aspic.

---

8. These letters were written to the prostitute Helen Jewett, shortly prior to her notorious murder by a client-*cum*-Significant Other. The story of Jewett and the context of her life and death are magnificently researched and recounted by Patricia Cline Cohen in the truly marvelous *Murder of Helen Jewett.*

Slavs to Fashion: Resistance to Linguistic Perestroika in Russia

Russian is another language where standardization and literacy congealed a language that had been transforming apace like a lava lamp. The establishment of a "standard" Russian was as geopolitically contingent and ultimately arbitrary as that of Standard English and French.

"Russian" as a Russian would recognize it begins as a tension between two poles. A thousand years ago, there was a strictly written variety called Old Church Slavonic, developed to write liturgical material and suffused with vocabulary and sense of prose style derived from Classical Greek. This was an ancestor to today's Bulgarian rather than Russian but was used throughout the Slavic-speaking world as a kind of lingua franca of writing, which was possible because the Slavic languages had not yet diverged as far from one another as they have since. Meanwhile, the closest thing to something considered "the best" *spoken* Russian was the mixture of dialects in Kiev, then the cultural capital but now contained within the region where what was once "Russian" is now termed the separate "language" Ukrainian.

In the 1300s, the center of power moved to Moscow, where the spoken variety was a mixture of the local dialect plus those spoken by newly arrived people from other Russian cities. In both Kiev and Moscow, the everyday language was highly variable, with writers, when they occasioned to transcribe spoken-style language on paper, free to choose among various alternates of a given word, with many "rules" of grammar varying according to the writer's regional background. The twentieth-century Russian philologist is struck by the "dialectalisms" rife in business documents or light-prose Russian of this period, as if Samuel Richardson's seminal novel *Pamela* were full of words like *yourn, critter,* and *a-fixin' ta.* Of course, the reason words and constructions like this pepper early Russian everyday text is that no single dialect had yet been chosen or composed as "The Language," such that what today is "dialect" was then just one among equals.

Through the centuries, some aspects of the Kiev and Moscow vernaculars had brushed off onto Old Church Slavonic while Old Church Slavonic materials made their way into the vernaculars as well. Yet by the end of the 1600s, the situation remained that, as one Russian grammarian at the time put it, *Razgovarivat'nado po-russki, a pisat'po-slavjanski* ("One converses in Russian but writes in Old Church Slavonic")–in other words, what was considered the "best" and "highest" Russian was a language no one actually spoke.

In the 1700s, however, as Peter the Great ushered Russia into richer contact with Western Europe and its intellectual traditions, the need arose for a "high," educated style of Russian to be used in conversation as well as on paper by the emerging intelligentsia. A rigid and liturgical variety such as Old Church Slavonic, based on Bulgarian rather than Russian itself, would do neither for bubbly salon talk and repartee in Moscow's mansions nor for a living creative Russian literature. Enter M. V. Lomonosov, who defined three "levels" of Russian of which even the highest was to be spoken as well as written. Even his version of "literary" Russian looks pretty peculiar to the modern Russian eye, however. It took N. M. Karamzin to create new learned words on Latin models to discourage the use of French ones: he created now-established words such as *razvitie* for *development*, which is composed of the same roots for *un-* (*raz-*) and *wind* (*vit'*)–a developing event can be said to be "unwinding," as the French *dé-velopper* was. Finally, Alexander Pushkin forged a dynamically expressive and flexible mixture of folk and elegant language to create what is today known as Standard Russian. Berlitz "Russian," then, only came into existence by the mid-1800s.

This standard, however, froze in amber a depiction of Russian that was in some ways already centuries out of date and that has since held back various natural developments in the language. Centuries before standardization, the sound *yeh* in accented syllables had become *yoh* after many consonants. Russian spelling caught up with this only at the end of the 1700s, symbolizing the new *yoh*'s as an *e* with an umlaut on it (*ë*). Thus the word for honey was once pronounced "myed" but is now pronounced "myoad," and is written мёд. The problem is that "educated" sentiment in Moscow preferred that a certain random collection of words "preserve" their original *yeh* pronunciation, just as English speakers are infected with an idea that *often*, which long ago morphed into "AW-fen," somehow "should" be pronounced "OFF-ten." Amid the standardization of modern Russian, these words were officially "set" in their earlier *yeh* forms, the result being various words where natural change would have left a *yoh* but "proper" pronunciation is nevertheless considered to be the old *yeh* form. Thus the word for *cross* today is "KRYEST" (крест), when if there had been no arbitrary standardization, it would long ago have become "KRYOST." This is even clearer given words of similar shape where the change was allowed to take its course, such as the word for *sparkle*, "BLYOST-ka," which began as "BLYEST-ka." The word for *sky* would be pronounced "NYO-ba" if Russian were spoken in a mountain valley and had never been written down, but was held back from its natural development and is thus still pronounced "NYEH-ba" (небо).

The words where the change was artificially blocked tended to be ones used a lot, as habit has a way of making old forms stick around in words so deeply ingrained in our consciousness. In English, we see this with the persistence of irregular plurals like *men, women,* and *children*. Along these lines, in some cases in Russian, the *yoh* change was retarded in a word's most common use but allowed to proceed in more marginal uses of the word that escaped being lassoed by the vagaries of oral fashion in midmillennial Moscow. The word for *sky* is also used for *palate* (makes sense!) and, in this usage, less likely to come up in conversation and thus more likely to escape the snares of "this year's fashions"; the word *is* "NYO-ba" (нёбо), having been allowed to take its natural course.[9]

Overgeneralization is attempting to play its hand in Russian in the realm of plurals. Russian has two main ways of forming the plural. For masculines and feminines, the markers are the suffixes *-i* and what is romanized as *-y* (the sound that comes out if you shape your mouth for *oo* and then keep your tongue in the same position but unround your lips): *bones* is *kost**i**, tables* is *stol**y**, books* is *knig**y***. For neuters, the suffix is *-a* or *-ja*: *words* is *slov**a**, fields* is *pol**ja***. For the past two hundred years, the neuter endings have been spreading slowly to many masculine nouns, being felt as the "default" plural ending: *officer* is a masculine whose "proper" plural is *ofitser**y***, but as often as not, one hears today *ofitser**a***; *years* is "supposed" to be *god**y***, but as often as not is *god**a***. What Russian "wants" to drift toward is a state where *-a* and *-ja* either cover a lot more ground than they do now or quite possibly become the plural end-

9. The only Cyrillic letters that correspond to Roman ones in sound are к, а, о, з, т, and м. This is perhaps why it has been said that Russian writing looks like Kaopectate written over and over again.

ing for both masculine and neuter nouns and just maybe eventually all nouns. No one objected when this happened with English *-s*, because the language had not yet been standardized and was thought of as a mainly oral language. Only the frozen representation of Russian on the page leads to a sense that the spread of the neuter plural is "uneducated" or, as a Russian friend's mother put it, how one speaks if one wants to sound like one grew up "in the village."

Vacuuming in Pearls: Writing Creates New Ways of Speaking

The transformation of First World speech is channeled into directions it would not have taken otherwise not only by widespread literacy, but by the stylistic tendencies of writing itself.

The Origins of Our Sense of "Educated Speech"

Adam Speaks Here is someone speaking English:

> Some of you say to me, "I'm not like you, I'm not a Congressman, I haven't got education, I haven't got work" . . . but you're a human being! And you know what you've got? You've got your hand, the power to use it to vote and use even those few cents you get from welfare to spend them only where you want to spend them!
>
> A young slave boy stood one day by the greatest ruler of his day, and God said to Moses, "What's in your hand?" Moses said only "I've got a stick." He said, "Well, let me use what's in your hand." God used that slave boy with a stick in his hand to divide the Red Sea, march through a wilderness, bring water out of rocks, manna from heaven, and bring his people to Freedomland.
>
> What's in your hand? *What's in your hand?* George Washington Carver was so frail he was traded for a broken-down horse as a slave boy; and George Washington Carver, sitting in his science laboratory at Tuskeegee told me, he said, "Dr. Powell," he said, "I just go out in the fields each morning at 5 o'clock, and I let God guide me, and I bring back these little things and work them over in my laboratory" and that man did more to revolutionize the agriculture

> science of peanuts, and of cotton, and of sweet potato, than any one human being in the field of agricultural science.
>
> What's in your hand? What's in your hand? Just let God use you, that's all. What's in your hand? I've got a string in my hand, that's all, and I'm flying a kite, and way up in the heavens, lightning strikes it and I, Benjamin Franklin, discover for the first time the possibilities of electricity, with a string in my hand. *What's in your hand?* Little hunchback sitting in a Roman jail, I don't have anything in my hand but an old quill pen but God says, "Write what I tell you to write!" Paul wrote, "I have run my race with patience, I have finished my course, I've kept the faith!"
>
> *What's in your hand?* Little boy, all I've got is a slingshot, and the enemies of my people are great and big and more numerous than we are. "Well, little David, go down to the brook and pick out a few stones, and come on back, and close your eyes if you want to and pull back that slingshot and let 'em go!" And David killed the biggest enemy, the leader of the giants against his people, and his people became free, just letting God guide a stone in his hand. And a few years passed and David is a king, and God says, "What's in your hand?" and he says, "I got a harp in my hand," and he says, "Well, David, play on your harp," and he played, "The Lord is my shepherd, I shall not want, taketh me to lie down in green pastures, leadeth me beside still waters, yea, though I've walked through the valley in the shadow of death I bear no evil."
>
> What's in your hand? Man hanging on a cross, "I've got two nails in my hand, Father, I stretch my hands to thee, no other help I know, if thou withdraw thyself from me, whither shall I go?" and that man, with two nails in his hands, split history in half, BC and AD. And what's in your hand tonight, people of Cambridge? You've got God in your hand, and he'll let you win, because he's on your side and one with God, always in the majority, so walk with him, and talk with him, and work with him, and stick together and fight together, and with God's hand in your hand, the victory will be accomplished, here, sooner than you dreamed, sooner than you hoped, sooner than you prayed for, sooner than you imagined. Good night and God Bless You.

That was the Reverend and Congressman Adam Clayton Powell, Jr., in the 1950s. It is a truly majestic speech, isn't it? And yet when you get right down to it, in stylistic terms it is quite simple. There is very little of a "written" or "fancy" nature in it. " 'Well, little David, go down to the brook and pick out a few stones, and come on back, and close your eyes if you want to and pull back that slingshot and let 'em go!' " The teacher of writing would suggest that the number of *and*'s be minimized, in favor of something more "compositionly" like " 'Well, little David, go down to the brook, pick out a few stones, and come on back; close your eyes if you want to, pull back that slingshot and let 'em go!' "

"I've got a string in my hand, that's all, and I'm flying a kite, and way up in the heavens, lightning strikes it and I, Benjamin Franklin, discover for the first time the possibilities of electricity, with a string in my hand." "Written" English would consider that extra "with a string in my hand" redundant.

"*What's in your hand?* Little boy, all I've got is a slingshot, and the enemies of my people are great and big and more numerous than we are." First, that redundancy again–we would be told to use one concise expression such as "our enemies outnumber us greatly" rather than the dramatic but somewhat frenzied and, technically, repetitious "the enemies of my people are great and big and more numerous than we are." And then, as Powell renders it, saying just "Little boy" sets the scene for us, and then narrative common sense and intonation tell us that the immediately following "all I've got is a slingshot" is what the little boy says. But in writing we would have to write, "There was a little boy," and then to indicate the boy's saying "All I've got is a slingshot," we would use "and he said . . ." or the like.

The rules of written English, then, would result in prose of this kind:

> I left for the fight full of a weird and violent depression, which I traced partly to fatigue–it had been a pretty grueling time–partly to the fact that I had bet more money than I should have–on Patterson–and partly to the fact that I had had a pretty definitive fight with someone with whom I had hoped to be friends.

That's James Baldwin ("Patterson" was boxer Floyd Patterson, and the fight was with Norman Mailer [big surprise]). That's beautiful English, too, but just a different kind, appropriate to its context, the printed page. *And* is used only once in the whole passage;

no redundancies; no "Little boy." Yet we would not want Powell to have phrased his speech in this style, anymore than we would want Baldwin to have expressed his ennui that night along the lines of "I left the house and I was really depressed, it felt almost weird; I was tired, for one thing, they had really put me through hell, and then I had bet all that money–Patterson–and I had had a big fight with this guy, I was hoping we were going to be friends, but that was it."

The obvious wholeness and legitimacy of Powell's utterance despite its flouting what we are taught as the "rules" of "good English" highlight that spoken language, rather than being a pale reflection of written language, is actually a distinct realm, with its own structures and aesthetic standards. The stylistics of Powell's speech are not symptoms of something afoot in America or the English language but represent the patterns that spoken language takes in all of the world's societies.

For example, when humans talk, they tend to partition units of information into brief packets (one analysis indicates that the average is seven words). These packets are typically run together with *and* or *so* or simply juxtaposed rather than linked and layered by strategies such as *Having had lunch, she proceeded to go to the ATM machine*–people just don't talk like this casually, if you think about it, regardless of education level. Also, in spoken language, a moderate degree of redundancy and repetition is a communicative plus, rendering dramatic contour and calling attention to the concepts the speaker wants to stress. For Powell not to have repeated "with a string in my hand" would have rendered the passage less effective, in not emphasizing that so much came from so little. In writing, the fact that Franklin accomplished all of this with just the string in his hand needs the repetition less, because we can see the business about the string still sitting up there on the line above. In speech, one can simply name one's subject–"Little boy"–and then jump right to something the boy says–"All I've got is a slingshot"–because gesture and intonation make the meanings clear. Writing requires us to be more explicit.

How does this relate to how some of the thousands of offshoots of the first language have developed? Today, in a literate society spoken and written language coexist side by side. However, it was not always so, and specifically, writing, in the chronological sense, emerged long, long after speech did: speech goes back 150,000-plus years; writing only about 6,000. Writing seems to us to be when language "came into its own" because that's when we can *see* it. But this is like the film career of the grand old actress Marie Dressler. Born in 1868, Dressler didn't attain film stardom (scattered silent appearances aside) until the late 1920s, when she was sixty; she was one of the highest-grossing and most beloved stars in Hollywood in the early 1930s but died in 1934. Dressler is beloved by her fans as a movie star immortalized on film, but her Hollywood years were actually only the last sliver of a career spent mostly on the stage, constituting not much larger a proportion of her life than 6,000 years constitutes that of human language. Her stage career seems like a "prelude" only because theatrical performances were not recorded–that is, the film career seems like "the real thing" only because we here today can *see* it.

In that light, humans worldwide were phrasing things along the lines of Adam Clayton Powell's "Little boy–'All I've got is a sling-shot'–and the enemies of my people are great and big and more numerous than we are" for tens of thousands of years before any language had been written down. This is how transcriptions of living speech in any unwritten language pattern, not like the Baldwin passage. In other words, expression of this kind was truly the language of, well, Adam.

Clearly, spoken language was not "waiting" for written conventions to come along and "tighten it up." Rather, written language conventions–and the tendency for educated speech to then mimic them–are an artificial add-on to human language, designed for the specific and highly historically contingent task of transcribing speech effectively into writing. Indeed, people heavily exposed to written language tend to acquire the capability of expressing themselves in this variety and use "written" strategies in their speech more often than people without education do. If you are reading this book, chances are that you could orally spin out a passage on the stylistic level of the Baldwin passage without great effort. This is what is known, in our "tall building" cultures, as "articulate" speech or "language skills." Yet restricting our sense of "articulate" to written-style language, inevitable though it is given the permanence and authority of the printed page, can obscure the following: if writing had never emerged, many of the stylistic constructions typical of written language throughout the world would never have acquired such a foothold in spoken language. Writing has a way of detouring human speech into customs that in the grand scheme of things are as unnatural as, well, reading itself, which has ruined the eyes of untold billions of humans, our eyes not having been designed for constant decoding of endless rows of tiny symbols.

Um . . . Why Couldn't They Write? The accessory nature of written conventions is especially clear when we note that, before English was standardized, when a writer had no established conventions of "written" English to refer to, even what was considered formal language displayed the very qualities we now associate with speaking rather than writing. In other words, there was not yet a variety akin to the style James Baldwin wrote in.

In early English texts, although we may feel "guilty" for saying so openly, we often get a sense of a certain clunkiness or choppiness in the stylistic construction of the prose–deep down we feel like "people didn't know how to write yet." And in a sense, they didn't, because there had not yet been established "a way to write." Thomas Malory's *Morte d'Arthur* of the late 1400s is a useful example. This passage has a certain majestic thrust, yet it would be hard to say that the flow of prose is as supple as that of Jane Austen:

> And thenne they putte on their helmes and departed / and recommaunded them all wholly unto the Quene / and there was wepynge and grete sorowe / Thenne the Quene departed in to her chamber / and helde her / that no man shold perceyue here grete sorowes / whanne syre Launcelot myst the quene / he wente tyl her chamber / And when she sawe him / she cryed aloude / O launcelot / launcelot ye haue bitrayed me / and putte me to the deth for to leue thus my lord / A madame I pray yow be not displeased / for I shall come ageyne as soone as I may with my worship

The slashes that the author uses as sole punctuation make even clearer the fact that this writing is based more directly on how people speak than we would expect of most written prose today; the slashes correspond rather well to the "idea packets," using linguist Wallace Chafe's term, into which people apportion their utterances. Like Powell, Malory prefers a simple *and* to mark each new occurrence rather than layering phrases together along the lines of "the Quene departed in to her chamber, holding herself such that no man should perceive. . . ." Like Powell, Malory jumps straight to Lancelot's saying "A madame I pray yow . . ." without specifying "Then Lancelot said, . . ." as if we were listening to Malory himself tell the story, assuming Lancelot's voice and demeanor to make clear that someone else is speaking.

This simply would not do in written English today, but it is only since the 1600s that this kind of writing is no longer accepted as appropriate for the page. By this time, English had been standardized and the Renaissance had encouraged a refashioning of English prose style on conventions set long ago for written Classical Latin. In other words, one no longer wrote as one talked or in the style of a bard reciting epic poetry in compact, memorizable strophes, and thus in his *Areopagitica* (1644), John Milton does not give us the impression of writing in a vaguely clumsy way:

> We should be wary therefore what persecution we raise against the living labours of public men, how we spill that seasoned life of man, preserved and stored up in books; since we see a kind of homicide may be thus committed, sometimes a martyrdom, and if it extend to the whole impression, a kind of massacre; whereof the execution ends not in the slaying of an elemental life, but strikes at that ethereal and fifth essence, the breath of reason itself, slays an immortality rather than a life.

This passage contains a large and nuanced thought within a single sentence by the manipulation of pronouns: in ". . . and if it extend to the whole impression," *it* refers not to the immediately preceding homicide and massacre but the persecution and spilling outlined in the last major proposition; "whereof the execution" lends a vividness and air of action about the already well chosen word *massacre*; and so on. The punctuation differs slightly from what we would expect, but Milton in no sense leaves us quietly thinking, "Why couldn't they write back then?"[10]

---

10. What remains unexplained for me, however, is why turn-of-the-century comic strips often leave you wondering why commercial artists "couldn't draw" yet. Although written language and its conventions simply did not exist until printing conditioned their invention, obviously, people of 1900 knew what good art was. Why, then, does it take until roughly the 1940s before most comic-strip characters are drawn in a streamlined manner pleasing to the eye? Yes, Little Nemo. But the early *Mutt and Jeff, Blondie, Bringing Up Father,* or *Popeye* look like they were drawn on cocktail napkins by drunks. Similarly, Mickey Mouse in his first two cartoons looks like something you'd kill on sight if you saw it, and the original Campbell's Soup kids would have *discouraged* me from eating anything.

The important thing, however, is that, to the extent that anyone can manage to talk in ways approximating written prose, this is in the strict sense a kind of trick, resulting from the fact that conventions of this sort are useful for writing and are thus exposed to us on such a regular basis that we internalize them. An analogy is bridge-and-tunnel gal Audrey in the musical *Little Shop of Horrors,* yearning to live the pat suburban life she has seen on 1950s and 1960s television sitcoms like *The Donna Reed Show* and unaware that the life depicted on these shows was itself a refracted iconography idealizing and distorting suburban life, rather than being a life that anyone had ever actually lived. At no point did housewives without significant psychosocial maladjustments vacuum their living rooms in elegant skirts and pearls, waiting to present a pretty picture to hubby when he got home. Similarly, in our tabula rasa state, it was not natural to consider tidily situated subordinate clauses or expressions like *with which to* and *without whom* as "the real language." When we try as much as possible to sound like a book, although in our lives this is socially necessarily in many contexts and many of us are so used to doing it that we barely have to think about it, the fact that this *is* necessary in Western cultures is based on an unintentional illusion that the conventions of writing are somehow "real" language.

"All Happy Families Are, Like, the Same, but Sad Families, They're All Sad All Kinds of Ways":[11] Russians Couldn't Write Either This historical trajectory was in no sense unique to English. A century and change before Malory wrote, written Russian tended to have the same air of informal discursiveness. In the 1300s, a passage in the *Primary Chronicles* doubts that Kiev was founded by a boatman along these lines:

> Ašče bo by perevoznik′ Kiy, to ne by xodil′ Tsajugorodu; no se Kiy knjažaše v rode svoem′; i prixodivšju emu ko tsarju, jakože skazajut′, jako veliku čest′ prijal′ ot′ tsarja, pri kotorom′ prixodiv′ tsari.

which translates as:

> For if Kiy was a ferryman, then he would not have gone to Constantinople; but behold Kiy was reigning in his tribe, and arriving at

11. The first sentence of *Anna Karenina* as translated by George W. Bush.

> the Emperor, as they say, that he received great honor from the Emperor, in the reign of which Emperor he came.

Jumping from one topic to another and not quite managing to get across a single coherent point, this passage, not at all atypical of literary style of the period, is almost reminiscent of children's speech ("Oh yeah? Yeah? Well, if he was just a ferryman then he wouldn't have gone to Constantinople, right? But no, he was the chief of the tribe and he went to the Emperor, like, the Emperor, and he got a big prize, when the Emperor was the big man up there, you know?"). Ultimately, however, have you ever tape-recorded yourself and your friends talking casually and then listened to it later? What is always striking is how few complete sentences of any length we actually tend to utter, how contrary our daily utterances are to the idealization of language we are constantly bombarded with on the page. We speak in "idea packets" or, better yet, when we try to spin out longer propositions, we risk being interrupted because our subconscious rules of discourse are founded on an expectation that people will talk in spurts. As a rule, in casual conversation human beings do not, and never have, spoken in tightly constructed sentences and carefully bounded paragraphs.

This is one thing distinguishing real life from plays, in which characters stand around making five-minute speeches while the other characters just sit and listen. If anyone does try to talk in chapters in real life, it's a little annoying–I once knew someone like this, and though the erudition and deathless zest for analysis were initially impressive and somewhat charming, it got old really fast. Passages like the Kiy passage, then, are simply the writing of a people who had yet to develop an artificial layer of unnaturally "anal" speech.

Again, today to a modern Russian, a passage as loosely constructed as the Kiy one looks almost like something somebody spray-painted on the side of a building. Russian's modern written register entails elaborately layered tapeworm sentences even farther from the spoken language than anything we can imagine in English, making even clearer that written style is a gloss on human expression, not its pinnacle. *The Master and Margarita,* a bizarre fantasy novel as seminal to Russians as *The Great Gatsby* is to Americans, is typical of literary Russian in being chock full of passages like this one:

> Ot fligelej v tylu dvortsa, gde raspoložilas′ prišedšaja s prokuratorom v Eršalaim pervaja kogorta Dvenadtsatovo Molnienosnovo legiona, zanosilo dymkom v kolonnadu čerez verxnjuju ploščadku sada . . .

The English version has this as:

> A haze of smoke was drifting towards the arcade across the upper courtyard of the garden, coming from the wing at the rear of the palace, the quarters of the first cohort of the XII legion; known as the "Lightning," it had been stationed in Jerusalem since the Procurator's arrival.

But in rendering the passage, the translator had to unravel the Russian phrasing, which word for word comes out as:

> From the wing at the rear of the palace, where lodged themselves the having-come with the Procurator to Jerusalem first cohort of the Twelfth Lightning legion, drifted smoke towards the colonnade across the upper courtyard of the garden.

Thus where the translator needs to give the background of the cohort in a separate phrase tacked onto the end of the passage, in Russian this "backstory," as actors call it, can be tucked in its entirety before the word *cohort* as a big mess of an adjective–thus just as one can say "the **large** cohort" or "the **victorious** cohort," one can say "the **having-come with the Procurator to Jerusalem** cohort." No Russian talks like this, but any literate Russian is casually familiar with sailing over constructions like this in newspapers and books. This "hyperadjective" tendency was imported into the emerging Russian standard modeled on a similar tendency in written German, a trait equally alien to any German conversation.

The Eskimo in the Gray Flannel Suit: When Indigenous Languages Go Written

Even today, when indigenous languages begin to be written, we can see new "written" traditions developing, packaging information into single sentences where the spoken variety preferred spreading things out over several short phrases. Somali has only been written regularly since 1972, and already, written Somali, instead of simply mirroring the spoken language, has taken on traits of its own. Here are two women talking animatedly about a piece of gossip; this is "the language" as it existed until 1972:

A: Wallaahi, dee, way iska fiicnayd–suurahay taqaan, haye?
swear um she just fine coyness-she knew huh?

I swear, she was just fine. She knew how to be coy, didn't she?

B: Waxaa iigu dambeysayba waa kaas. Waxay iigu
what for-me was-last-time that what-she for-me

darnayd ayaantay
was-the-worst the-day-she

The last time I saw her was then. The worst thing for me was the day she said

Amina ku tidhi "Ninkayga, ninkayga . . ."
Amina to she-said my-man my-man

to Amina, "My man, my man . . ."

A: "Ninkaygay igu dirayaan."
my-man-they to-me they-set

"They're setting my man against me."

B. Adduunka, kelmaddaasi weli waa xasuustaa, ka warran!
the-world that-word still I remember how-about-that

Imagine! I still remember that word, how about that?

A. Dee, horta waa runoo waan ku dirnee, ma og tahay?
Um the-first truth-and we to send-and know being

Taasi ma been baa?
that lie

Uh, first of all it's true, we did set her against him, don't you know? Is that a lie?

One thing marking this excerpt as spoken rather than written language is its signs that the speakers are subjectively engaged in the topic rather than analyzing it from "on high." *I swear, didn't she?, Imagine!, Don't you know?, Is that a lie?*–these interjections arise from a heart-level engagement with the subject and would be unlikely in a passage written about the event in the third person. There are also, although this is unclear to someone who doesn't happen to speak Somali, contractions in this passage: the word on the first line *suurahay* is actually composed of three words *suuraha* "(the) coyness," a word *ayaa*, which lends a kind of affective emphasis in the "I'll bet!" sense, and *ay* "she."

Things are quite different in this excerpt from a Somali newspaper editorial:

Isbeddelkaasi	ka	dhacay	Boortuqiiska	uu	yahay	mid	hore	u
self-change-that	from	happened	Portugal-the	it	being	one	forward	to

wadi	doono
take	will

The self-transformation that happened in Portugal is one that will continue

siyaasadda	isticmaariga	ah.	Hase	yeeshee	waxa	dhici	karta
policy-the	colonialism-the	is	let	be-but	what	happen	can

inuu	la
that-it	with

the policy, which is colonialism. However, what can happen is that it

yimaado	tabo	cusub	uu	ula	jeedo	inuu	waqti	ku	kasbo . . .
come	tactics	new	it	to-with	mean	that-it	time	with	benefit

would come with new tactics whose purpose is to gain time.

No *Ain't it the truth?* or *Tell me I'm wrong!* and contractions are fewer and farther between. In addition, the editorial makes extensive use of relative clauses. In English, we can say *The man* ***who told the officer what had happened*** *was a villager with no money,* but in casual speech, this would be more likely to come out as something like *This villager with no money, he told the officer what happened.* Relative clauses exert more of a processing load on us when we are speaking or listening "on-line" than when we are reading text and can ingest more verbiage in a swallow. As a result, in the editorial there are several relative clauses, "the transformation that happened in Portugal" is "one that will continue the policy, which is colonialism"; news tactics "whose purpose is to . . ."

It is not that relative clauses did not exist previously in Somali but that they were not used as frequently in speech; in written Somali, what began as a side dish has become a staple. Furthermore, Somalis are now using more of these new written conventions in more formal registers of speech, such as in lectures, speeches, and meetings.

In the Eskimo language Inuktitut, as it has come to be written in newspapers, relative clauses are actually developing from the ground up, having not existed at all in the language as it was spoken by hunter-gatherers. In spoken Inuktitut, the closest equivalent to English *The Inuit who are council members* would have been to say what would translate literally as "The council-membering Inuit." Today, however, the word for "this," *taana,* has developed a new usage as a relative pronoun:

Taimataq	uqalauqpug	Luis Amalat,	taana	ilinniarvigjuarmiutaq
so	spoke	Louis Hamelin	this	university-member

silaturiarvingmi	atilingmi	lavalmi	kapaak	sitimi.
in-nature-research	in-one-called	Laval	Québec	in-city

. . . says Louis Hamelin, who is a biology professor at Laval University in Québec City.

The entire repertoire of the language Inuktitut, then, now includes a construction unknown before it was written, and it is likely that relative clauses will begin to appear in the spoken language among speakers using it in formal functions.

When the English, Russian, Somali, and Inuktitut cases are taken together, it becomes clear that heavy use of relative and subordinate clauses is largely an artificial decoration on "natural" human speech, which largely uses them sparingly. Natural speech began preferring *The man, he went up there* and *She knows it–she saw it!* and often still does. Some linguists have even speculated that originally all humans spoke largely with just main clauses in this fashion, and this is manifest in the Hebrew Bible and the Greek of Homer, in both of which, despite their obvious brilliance and grandeur, relative and subordinate clauses are a sometime thing at best. It has even been hypothesized that Proto-Indo-European had no subordinate clauses, which would have been predictable of a language that was never written.

Nursery Rhymes versus Jacques Cousteau: Formal Language versus Written Language

That last sentence requires a bit of comment and clarification. These developments must not be taken to imply that languages that have not developed writing conventions remain an evolutionary step behind the mostly Western ones that have. For one, throughout this book we have seen that languages spoken by hunter-gatherer societies or uneducated peoples tend to be dripping with a degree of historically accreted complexity that makes English look like Esperanto. All those needless gender markers and inflections, the Tuyuca evidential markers, the barely predictable plurals of Luo, the six tones in the Cantonese of even the humblest Chinese peasant–none of them have anything to do with relative and subordinate clauses, and they are integral to even the most basic statements (greetings, exclamations, drunken grunts, children's songs). In addition, even "indigenous" societies often retain a special level of language for ritual purposes that is more elaborated than the spoken language, differing enough from everyday speech that it must be consciously learned. When ethnobotanist Mark Plotkin spent time with the Tirió Indians of Surinam, he noted that the Tirió language used for telling legends was more complex than the language ordinarily spoken, and this is true of ritual language in many preliterate societies.

This last differs from writing conventions, however, in that ritual speech often differs from the spoken in retaining features that *were* part of the spoken language centuries earlier, the features having persisted in the codified telling of tales while they naturally morphed away in the spoken variety. An analogy is English's *Once upon a time*–a rather queer little phrase if you think about it; what's "upon a time"? Or any number of phrases from Mother Goose Nursery Rhymes: What in the world does "Bye Baby Bunting" mean?[12] Russian folktales, too, are scattered with odd little expressions dating from earlier stages of the language: their *Once upon a time* is *Žili-byli*, which translates today as the nonsensical "Lived were" but arose when Russian had a pluperfect tense of which this was an ordinary living example.

However, in indigenous ritual languages this degree of remove from the modern language permeates the entire speech style, rather as if the Nursery Rhymes were in Middle English or Russian grandmothers told the old folktales in Old Church Slavonic. In reference to Surinam again, many of the first slaves transported to Jamaica were Sranan-speaking slaves brought from Surinam, and Sranan was a foundation element in the mixture of languages that eventually became Jamaican Creole. Today, descendants of slaves who escaped Jamaican plantations and settled in the interior use a ritual language in ceremonies that turns out on examination to be a form of Sranan, not spoken by any Jamaicans themselves anymore but, having been the foundation stage in Jamaican patois's history, naturally retained as the archaic ritual variety.

Writing conventions, on the other hand, create a brand-new variety of linguistic expression no one ever used before in casual speech, or they bring to the foreground constructions that in natural speech were once only one among many. Written-language styles, from this perspective, take their place as another of the many aspects of First World life deeply ingrained in our consciousness as inextricable from humanity and yet ultimately mere epiphenomena of technology.

In illustrations of marine life before the late 1800s, sea creatures are almost always drawn washed up on shore, decoratively adorning the frame of the plate in an abstract manner or, if they are actually in the water, popping their heads up above the surface as seen from the shore. William Gladstone and Abraham Lincoln did not page through books about the sea full of pictures of marine creatures actually swimming around in their underwater habitat the

---

12. And what was "curds and whey"? Curds of what? Look up both *curd* and *whey* and then imagine someone eating a room-temperature bowl of this with relish (*curd* is a variant of the word *crud*!). Or Jack Horner pulling a plum out of a pie–why did he stick his thumb in rather than his pointer finger? Presumably just to rhyme with *plum*–but then what was a plum pie and how good could that have been? And why did dismembering the pie make him a good boy?

way we do. Why? It's so simple it's almost embarrassing, and yet we'd never think of it–back then, *How could anyone have ever seen marine animals living in their natural habitat?* Diving and snorkeling technology were virtually nil, and to the extent that we can hold our breath and nose around underwater near a shore, moving water is generally hard to see through and the "funky stuff" lives farther out anyway. If we duck our heads underwater on the Atlantic coastline, for example, we do not see a colorful panorama of flying nautiluses, coral, and angelfish. Only after a home aquarium craze in England did "underwater" views of sea creatures begin to take hold. To George Washington or George the Third, however, a fish was something dead on a plate. Even now only a small subset of the population has snorkeled, and an even smaller one has dived in the deep sea–our spontaneous mental image of squid and sea anemones doing the vaguely disgusting things they do in their natural habitat is due directly to the technology that exposes us to aquariums, deep-sea photographs, and nature specials.

In the same way, our sense that English "at its best" is the likes of

> Most efforts of psychologists within this tradition have been concentrated on looking for evidence demonstrating the psychological reality of one or another hypothesized underlying representation or of one or another type of linguistic rule.

is a by-product of technology; specifically, tendencies arising when spoken language is transcribed onto the page to be taken in by the eye rather than the ear.

Spoken Language: The Unsung Marvel

Make no mistake: I love written language deeply and enjoy few things more than composing prose on the page. Yet our natural sense as First Worlders is that written conventions are "improvements," refinements on an initially chaotic, "unfit" conversational style. Books on the standardization of Western languages in past centuries typically describe the process as "improving" the tongue. This, however, implies that there is something "backward" about the way most people on earth speak, because most languages are not regularly written. Properly, writing conventions are useful, perhaps aesthetically savory, but ultimately accessory additions to a language that was quite complex and nuanced in its unwritten state. In other words, writing conventions are but an art. The manipulation of the basic materials of a language spoken by the homeliest, most isolated group of humans is, on examination, as much of a miracle as Tolstoy's crafting of the prose in *Anna Karenina*.

Here are various things people in my life, all of them members of the homely and isolated culture known as America, have said spontaneously over the years that have doubled me over with laughter and delighted me with the richness of ordinary human use of language:

> (On a urine test) "So the first time I went they gave me a little tube and they said "Here, fill this up" and so that was okay, but then the next time I went they gave me a *jar*–I'm afraid the next time I go they're gonna give me a bucket!"
>
> "Her teeth look like a map of Paris!"
>
> John: I tried to help, and what did I get for it? Villainy and debasement.
> Friend: Who put you in de basement?
>
> (About a golden-age female celebrity of a certain age and avoirdupois) "Oh, the last time she saw her toes was in '62. And the last time she *saw* was in '67. And the last *time* . . . well, we won't talk about that."

And's all over the place, nary a relative clause dares raise its head, and yet these passages, in my view, are every bit as much an art as the *Areopagitica*.

The latter, in all of its glory, represents in the strict sense a perversion of what human language evolved to be. Analogized to biological evolution, standardized languages spoken by literate populations are living fossils, which have then had their genes adjusted for technologically advantageous purposes. The standard dialects of the Fortune 500 of the world's languages, then, are genetically altered coelacanths.

7

Most of the World's Languages Went Extinct

This book has been dedicated to an analogy between biological evolution and human language. Like animals and plants, languages change, split into subvarieties, hybridize, revivify, evolve functionless features, and can even be genetically altered. The analogy continues in that languages, like animals and plants, can go extinct.

As animals and plants drive one another to extinction by nosing one another out of ecological niches in competition for sustenance, in the past languages have usually gone extinct when one group conquers another or when a group opts for a language that it perceives as affording it greater access to resources it perceives as necessary to survival. Typically, a generation of speakers of a language becomes bilingual in one spoken by a group that is politically dominant or endowed with valuable goods or access to same. This bilingual situation can persist across several generations, but as often as not, the inevitable tendency for languages to be indexed to social evaluations takes its toll. Usually, through time new generations come to associate the outside language with status and upward mobility and the indigenous one with "backwardness." This is especially the case when the dominant language is a First World "tall building" language associated with money, technology, and enshrinement in the media while the indigenous one is an obscure tongue spoken only in villages.

A point arrives when one generation speaks the outside language better than the indigenous language, largely using the latter to speak with older relatives and in ritual functions. As such, these people do not speak the indigenous language much better than many Americans might speak French or Spanish after a few years of lessons in high school. One is unlikely to speak to one's child in a language one is not fully comfortable in and does not consider an expression of oneself. It is here that a language dies, because a language can only be passed on intact as a mother tongue to children. Once it is spoken only by adults and is no longer being passed on to children, even though it will be "spoken" in the strict sense for another several decades, it will die with its last fluent speakers.

Our natural sense is to suppose that, as long as the language has been written down or codified in a grammar, then it need not be dead forever. However, grammar writing is a relatively recent practice, and in the absence of a grammar, a dead language's full apparatus is only evident when there is a considerable volume of writings. This is in turn only the case for a small number of "big actors" such as Latin. Because writing itself is a relatively recent invention–as it has been put, if humans had existed for just one day, then writing would have been invented about 11:00 P.M.–obviously even these potential paths of rescue have been unavailable to human language for most of its existence. Until 11:00 P.M., once a language went extinct, it was gone forever.

An extinct language before the advent of writing is even more unrecoverable than an extinct life form. Life forms may leave their impressions as fossils, and technology gets ever closer to allowing us to someday at least partially resurrect ancient life forms through remains of their DNA. However, a language could not leave an "imprint" before writing existed, because an individual language is not encoded in a person's genes. If the ability to speak is genetically encoded, we can be quite sure that this inheritance is a generalized one allowing someone to speak any language on earth. The particular word shapes, grammatical configurations, and various irregularities that characterize any one language are the result of largely random accretions through the millennia, no more reproducible from basic human materials than the form of an individual snowflake is from the water droplets that it began as.

And even when a language is preserved in writing, there is a long trip indeed between the tales, recipes, battle accounts, and poetry preserved on the page and the language being used daily by living, breathing human beings as an expression of their souls.

Many of us can attest to this from our exposure to Latin–no matter how good you may have gotten at those declensions, conjugations, and ablative absolutes, even this was a long way from speaking the language fluently, and what life conditions can we even imagine, outside of the clergy, where fluent Latin would be natural or necessary? Languages die when others take their place–we don't *need* Latin or any dead language, because we've got languages of our own. As often as not, a revived language hovers in the realm of the "undead"–part of the revivification effort entails gamely making space for the language in lives already quite full without it and sometimes even vaguely discomfited by its return.

World History: A Treadmill to Linguistic Oblivion?

It Was Ever Thus–To an Extent

Like biological extinctions, language death has been a regular and unsung occurrence throughout human history. We have records of Indo-European languages now no longer spoken, such as Hittite from present-day Turkey and Syria, and Tocharian, spoken by Europeans who penetrated as far east as present-day China. There was once a Romance language spoken on the Adriatic shores called Dalmatian, a kind of transition between Italian and Romanian, whose last speaker died in 1898.[1] The Romance languages in general spread in a continuous patch from Portugal eastward until a break after Italy, turning up again only in Romania, with the exception of some dots of odd Romanian dialects spoken in the interim. As was Dalmatian, these dots are remnants of what once were many other Romance languages filling in today's gap–languages that died in the face of encroaching Slavic varieties now spoken in the former Yugoslavia. A Slavic language called Sorbian (or Wendish) is spoken within German borders and, predictably, has lost in the competition with German and is now only spoken by a few elderly people. King Arthur represented the Celtic peoples

1. Unfortunately he died toothless, rendering the data elicited from him somewhat fuzzy around the edges–particularly awkward because he was the only source of the language ever recorded.

who once inhabited all of the British Isles and significant swaths of Iberia and present-day France. The onslaught of the Romans and then the Vikings pushed the languages they spoke to the margins. Gaelic hangs on tenuously in Ireland, as does an offshoot variety in Scotland; Welsh does so in Wales; and Breton is fighting for life in northwestern France. But the Gaulish that the *Asterix* characters are supposed to be speaking has not been heard from since about A.D. 500, the last full speaker of Cornish of Cornwall died in 1777, and Manx of the Isle of Man died in 1974.

In other cases, language deaths in the past are only reconstructable by inference. In Africa, where languages often change from one small region to another, the Maa language is relatively unusual in being spoken across a belt of territory incorporating two countries, stretching from the top of Kenya to the middle of Tanzania. Peculiarities among various groups of its speakers attest to Maa having "killed" local languages in its spread, the Maa being a traditionally successful pastoral people who migrated widely in the past in search of grazing lands.

In northern Kenya, there are Maa speakers who stand out in being hunter-gatherers instead of pastoralists. These Dorobo peoples assist the Maa of the area in their herding, and the Maa's oral tradition mentions having met hunter-gatherers in the past. Presumably before the meeting, the Dorobo spoke their own language, this made even more likely by the cultural distinctiveness they retain today. Southward there are other Maa speakers whose cultural distinctiveness tips us off to language death in days of yore. The Camus people on Lake Baringo of Kenya farm and fish, traits alien to and even looked down on by traditional Maa peoples; the Arusa of Tanzania also remain farmers, though speaking Maa. In other cases, the death of languages in the face of Maa is concretely visible as speakers remain who speak a shredded version of the original language: the Elmolo, Yaaku, and Omotik languages are now only spoken by the very old, their communities having opted for Maa in connection with the benefits of the pastoral life style.

It was ever thus, then. No more ammonites, *Pteranodon*, or eighteen-foot-tall rhinoceroses;[2] no more Hittite, Dalmatian, or Elmolo. There is a sense in which we cannot help but regret the demise of any of the endlessly marvelous permutations of life or

2. Yes, there *were!* Just imagine that.

language, and surely the demise of each creature or language is in the strict sense a tale of marginalization and erasure. Animals and plants have vanished as often in catastrophic grand extinctions as through gradual outnumbering by more successful competitors, and languages have often died as the result of violent conquest, enslavement, and oppression. However, under ordinary conditions, we could perhaps congratulate ourselves that naked conquest of this sort is no longer officially sanctioned by the world community (even if, sadly, conditions too often leave such events to be allowed passively to proceed, especially when the people in question are not perceived as commercially important) and that our enlightened awareness of the value of diversity combined with the availability of writing will further help ensure that languages will no longer disappear at nearly the rate that they used to.

This, however, is an understandable but mistaken view. Our parallel with animals and plants unfortunately extends to the fact that, today, languages are in fact disappearing at a rate as alarmingly rapid as that of flora and fauna. The same geopolitical forces that are raping the global environment are also vaporizing not just the occasional obscure tongue spoken in remote regions, but most of the world's six thousand languages. Today, a subset of the "top twenty" languages (Chinese, English, Spanish, Hindi, Arabic, Bengali, Russian, Portuguese, Japanese, German, French, Punjabi, Javanese, Bihari, Italian, Korean, Telugu, Tamil, Marathi, and Vietnamese)[3] are imposed as languages of education and wider commerce throughout the world. The result is that ninety-six percent of the world's population speaks one or more of these top twenty; that is, these people speak one of these languages in addition to an indigenous one, and there is a threat that succeeding generations will learn only the dominant one and let the indigenous one die. This means that only *four percent* of the world's population is living and dying speaking *only* an indigenous language.

This imbalance of power leads to some rather gruesome predictions. By one reasonable estimate, ninety percent of the world's languages will be dead by 2100–that is, about fifty-five hundred full, living languages will no longer be spoken about 1,125 months

3. Notice that India is so populous that languages many of us have never heard of are spoken by more people than almost any others in the world (Bihari, Telugu, Marathi).

from when you are reading this. As David Crystal puts it, this means that a language is dying roughly every two weeks.

Many of the languages we are most exposed to are among the top twenty or will be among the five-hundred-odd "medium" languages that will likely survive the impending mass extinction (Catalan, Finnish, Wolof, Thai, Tagalog, etc.), and all of these languages have been so richly documented in writing and in recordings that, even if they lost all of their native speakers, their revival, or at least maintenance on life support, would be at least technically feasible. Thus it can be difficult to appreciate the massive loss that more widespread language death will entail.

The Native American situation is illustrative. Before the arrival of Europeans, there were about three hundred separate languages spoken by Native Americans in what is today the continental United States. Today, a third of those languages are no longer spoken, whereas all but a handful of the rest are spoken only by the very old and will surely be extinct within a decade or so. The current situation is as if, in Europe, Albanian, Frisian, Romanian, Basque, Catalan, Occitan, Welsh, Lithuanian, Latvian, and Irish and Scottish Gaelic were no longer spoken, and meanwhile only English, German, and Russian were still being passed on to children, with Swedish, Norwegian, Danish, Icelandic, Dutch, French, Spanish, Italian, Portuguese, Serbo-Croatian, Macedonian, Polish, Bulgarian, Finnish, Estonian, and Hungarian only spoken by very old people, viewed as "quaint" and backward by young people jetting around in sports cars.

Each of these Native American languages was an astoundingly complex and remarkably beautiful conglomeration, presenting the glorious kinds of baroquenesses we have seen in Cree. Europe is covered mostly by languages of one family such that all are based on a common general "game plan," but the Native American languages spoken north of Mexico constituted at the very least two dozen families, with a range of variation across the continent as broad as that on the entire Eurasian landmass, taking in Indo-European, Chinese, Japanese, Arabic, and others.

The First Crack in the Dam: The Neolithic Revolution

The trend toward a decrease in the number of the world's languages is, in large view, not an isolated phenomenon but one symptom of general trends in human development in the past several millennia. Until just about eleven thousand years ago, humanity worldwide consisted of relatively small groups of hunter-gatherers. This life style was not one inherently geared toward population increase and spread, and thus the world was feasibly shareable by large numbers of such groups, with minimum occasion for one group to exterminate another one along with its language. We can be sure such things happened but generally on a very local scale, counterbalanced by the birth of new languages as groups that reached a certain size spawned offshoot groups who moved away from the original group, their speech eventually developing into a new language. It has been estimated that this world could have harbored as many as one hundred thousand languages, and the scenario has been termed *linguistic equilibrium.*

Large-scale language death began with the development of agriculture in many societies, starting in about 9000 B.C. Agriculture required large expanses of land, and its greater yield of food led to hitherto unknown population growth. Cultivation allows the amassing of food surpluses, which, freeing certain classes of people from hand-to-mouth subsistence, is the basis of the development of hierarchies of specialization that breed technological advances. Armed with these, and in constant need of extra space as their populations burgeoned, agricultural societies quickly began overrunning hunter-gatherer groups worldwide.

Even the way in which the world's language families are distributed today makes it clear that language death has been a regular part of human existence for several millennia. In India, roughly speaking, the languages spoken in the top half are Indo-European languages such as Hindi, Bengali, and Marathi, whereas in the bottom half, languages of another family called Dravidian are spoken, including Tamil, Kannada, Telugu, and Malayalam. However, the subdivision of space is not perfect: there is the occasional Dravidian language spoken way up in northern India or even as far northwest as present-day Pakistan. What are those people doing way up there? From our present-day perspective, it looks as if certain Dravidian speakers decided at some point to pack up and move thousands of miles away from their homelands. Much more likely, and supported archaeologically, is that Indo-European speakers slowly moved southward, with formerly spoken Dravidian languages dying along the way as their speakers were incorporated into the invaders' societies for generations and gave up their original languages. Life is never tidy, and naturally some pockets remain that the invaders never happened to get to. Thus the Dravidian "outliers" are remnants of a once greater variety of Dravidian languages spoken in southern Asia.

A similar case is the Dahalo language of Kenya, unusual in having the clicks otherwise found only way down in the south of Africa, among a small group of languages called Khoi-San and a few Bantu languages spoken near them such as Xhosa and Zulu. It is easy to see why the Bantu languages have the clicks–language contact long ago with Khoi-San speakers. However, what are clicks doing way up in Kenya? Clicks are so extremely rare cross-linguistically–otherwise found only in one Australian language, and even there, only in a special "secret" variety of that language–that it is unlikely that the clicks developed in Kenya merely by chance. Most likely, Khoi-San "click" languages were once spoken more widely in Africa, and Dahalo is one of the only remnants of that situation. What this means is that untold numbers of click languages must have died as Bantu and other peoples spread into what began as click-language territory. Again, archaeological evidence supports this scenario: skulls of people of the ethnicity who today speak Khoi-San languages have been found as far north as Zambia, and the Bantu takeover apparently occurred within a mere few centuries' time after 1000 B.C.

In the New World and Australia, Europeans similarly overwhelmed Native American and Aboriginal languages, assisted by the germs that living among livestock had immunized them to but that quite often decimated indigenous hunter-gatherer populations on impact.

Situation Critical: The Downsides of the Global Economy

Thus even today's six thousand languages constitute a vast decrease in the number of languages that existed before the Neolithic revolution. Today, however, a second revolution, which some leftist political commentators term the imperialist one, is having an even starker effect on how many languages are spoken in the world.

During the Neolithic revolution, when a language spread across an area, it generally did so relatively slowly such that, by the time the spread was complete, the language had already developed into several new ones, which continued to spawn new ones in turn. For example, by the time Latin was disseminated throughout the Roman Empire, its progenitor Proto-Indo-European had elsewhere in Europe already split into several branches such as Germanic, Slavic, Celtic, Hellenic (Greek). Then Latin itself developed into more than a dozen new languages, the Romance languages, while at the same time Proto-Slavic was developing into several new tongues. Thus, though Europe was once covered by languages now lost forever, this original diversity was replaced at least partly by new diversity. Furthermore, until recently, Europeans were unable to physically take over tropical and subtropical regions, where farming methods developed for temperate climates were ineffective and diseases Europeans had no immunity to tended to kill them, just as their own diseases tended to kill Native Americans and Australian Aborigines.

However, in the past few hundred years, the development of capitalism and the Industrial Revolution, and its resultant technological advances and encouragement of strongly centralized nationalist governments, have led a certain handful of languages to begin gradually elbowing not just many but *most* of the world's remaining languages out of existence. The urgencies of capitalism require governments to exact as much work and allegiance from their populations as possible, and the imposition of a single language has traditionally been seen as critical to this goal, especially within the nationalist models that have ruled since the 1700s. Recall the active hostility of the French government to the Occitan dialects and other "patois" of France, in favor of a scenario under which everyone in France spoke French.

In our era, climatological boundaries present few barriers to the onslaught. Today, language death is often caused less by physical conquest than by gradually yoking indigenous peoples into a centralized cash economy. This is often done by transforming their traditional life styles on site according to what their local topography can bear, with the aid of advances in agricultural technology. In other cases, the dominant power renders much of the population migrant laborers, spending half of their lives working in cities, this facilitated by modern transportation technology (assembling part-time work forces drawn from afar was more difficult before the invention of trains, for example). In the past few centuries, a great many human societies have been drawn from independent subsistence on the land into dependent relationships with capitalist superstructures, with traditional ways of life often actively discouraged in favor of new practices geared toward supplying the central government with salable resources.

A Skeleton of Its Former Self: A Language Withers Away

What happens to a language as it dies? Generally, the last generation of fluent speakers has learned it only partly, never truly living in the language, using it only in the corners of their lives. As a result, the language is slightly pidginized. However, whereas in many cases a pidgin has been a temporary "setback" on the way to its expansion into a new language, the moribund variety of a dying language is a step along the way to permanent demise.

"I Wish I Had the Words": Atrophied Vocabulary

Just as pidgins such as early Tok Pisin had restricted vocabularies, dying languages' vocabularies are constricted, with many single words pinch-hitting for concepts that were expressed by several more specific ones in the living language. Cayuga is a Native American language originally spoken in New York State. Under the Jackson Administration in the 1830s, as *you was* popped up in letters written by white clerks in New York City, many Native Americans were relocated to Oklahoma, intended as a delineated Indian Territory, and Cayugas were among them. By 1980, only a few elderly people spoke any Cayuga but had been thoroughly English-dominant all of their lives, and their Cayuga was seriously frayed around the edges as a result. The Oklahoma Cayuga had a word for *leg* but none for *thigh*, a word for *foot* but none for *ankle* or *toe*, words for *face* and *eye* but none for *cheek* or *eyebrow.* Where full Cayuga had a word specifically meaning *enter*, these old people substituted the more general word *go*, such that *Come into the house* was rendered as *Go into the house.* The nuance of where the speaker was in relation to the house–determining whether one would say from the porch ***Come** in* or say from a hill up yonder ***Go** in*–was left to context.

The Genericization of a Language: The Demise of the "Hard Stuff"

Just as pidgins strip away aspects of language not necessary to basic communication, dying languages are marked by a tendency to let drop many of the accreted "frills" languages drift into developing through time. In a language that one uses little, the first thing to start wearing away in the grammar is, predictably, the "hard stuff" that takes lifelong daily practice to learn and retain.

The Death of Inflections One "frill" in a language is the inflectional prefix or suffix, which quite a few languages do without. Inflections arise accidentally through time from what begin usually as separate words, which as inflections become one of the challenging aspects of language to learn, entailing lists of arbitrarily shaped bits of stuff signaling concepts such as gender, person, and number that are in any case either clear from context or unnecessary to communication. People who use an inflected language day in and day out learn the inflections with ease and have no trouble retaining them throughout their lives; they become as ingrained as walking. But in dying languages speakers have often never mastered the inflections fully or have lost control of them in time, and thus to them the inflections become "hard," just as they would be to a foreigner learner.

Thus speakers of a dying inflected language often avoid using inflections in favor of more immediately transparent constructions, just as an English speaker feels as if he has gotten a kind of break when finding out that, in Spanish, instead of *hablaré* "I will speak" he can also say *Voy a hablar*, which allows him to get around calling up the ending in favor of using a form of the verb *go* that he has already learned. In traditional forms of Pipil, spoken in Guatemala and El Salvador, there were future inflections, such that *I will pass* was:

Ni-panu-**s**
I- pass- will

But today we mostly see this in old texts; elderly living speakers might cough it up for money, but in the Pipil they speak, stripped down in comparison with the living language of yore, the future is expressed with a "going to" construction. *I'm going to do it* is:

Ni-yu ni-k-chiwa
I- go I- it-do

Thus Pipil has moved in the direction of pidginhood, paralleling the tendency in pidgins to express the future with "going" expressions (although many old languages happen to express the future this way as well).

The Soul of Celtic Melts Away In Welsh, by way of review, the first consonant of a noun often changes, depending on which posessive word comes before. The word for *cat* is *cath*, but the word takes on a different form as used with the word for *my*:

eu cath "their cat"
*fy **ngh**ath* "my cat"

The *his* and *her* case is particularly interesting because the same word is used for both: only the change in the consonant shows whether *his* or *her* is intended in the meaning:

*ei **g**ath* "his cat"
*ei **ch**ath* "her cat"

In Welsh's relative Gaelic–namely, the Scottish variety spoken in Sutherland County–speakers in their forties were the last generation of fluent speakers left in the early 1970s. Gaelic has the same kinds of consonant changes as Welsh does, and one sign of the decay of the Sutherland speakers' Gaelic was the gradual breakdown of these rules. To say *She was kept in* in living Scottish Gaelic, laid as all Celtic languages on a basic foundation quite different from English's, one says literally "Was she on her keeping in":

Bha i air a cùmail.
was she on her keeping-in

This is a little challenging to wrap our heads around, but for our purposes concentrate on the last two words:

Bha i air **a** **cùmail**.
was she on **her keeping-in**

To say ***He was kept in***–that is, "Was he on his keeping in"–one uses the same word as for *her*, but the consonant in the following verb changes:

Bha e air a **ch**ùmail.
was he on his keeping-in

And to say ***They** were kept in*, there is a different consonant change:

Bha iad air an **g**ùmail.
was they on their keeping-in

In the moribund Scottish Gaelic of Sutherland County, however, as often as not, in all three cases the form ***ch**ùmail*, properly used with *his*, was used for all three:

Bha i air a **ch**ùmail.	"She was kept in."
Bha e air a **ch**ùmail.	"He was kept in."
Bha iad air an **ch**ùmail.	"They were kept in."

These speakers have a general sense that there is some consonant change after possessive words but have not mastered the particular changes that each possessive pronoun requires or does not require. Thus just as we might do in trying to learn to speak Scottish Gaelic, these speakers simply generalized one kind of change to all persons.

There's Speaking and There's Speaking The last generation to speak a language is often incapable of being *articulate* in it as well, a crucial indication that the language is no longer capable of expressing full humanity.

There are scattered examples in English of concepts that are expressed as a single word incorporating both the object and the verb together: *He sat the baby for her* is more often rendered as *He babysat for her.* In many Native American languages, however, this process is central to basic expression, usable for just about any commonly occurring verb–object combination. In Cayuga, to say *She has a big house,* one might say "It house-bigs her," in the sense of "Things have it that she has a big house." Moreover, all of this is one word: *Konǫhsowá:neh.* When to use expressions like this and when not to are central to manipulating language artfully in these cultures, in the same vein as word choice and relative clauses are for us.

Of course, as we in particular know, a language can do just fine without this sort of thing, which evolves accidentally in certain languages through time–earlier we saw it having done so just in the past century with "camp-sat" in Ngan'gityemerri in Australia. It's an extra and, as such, one of the first things to start wearing away as a language containing it dies. In living Cayuga, to render *She has a big onion* within a narrative, one would likely say "It onion-bigs her." In dying Cayuga, however, speakers are more likely to just say something like *The onion is big* or *Her onion is big.*

When a language is dying, then, its last speakers typically render it in the very way that we or another foreigner might, taking the easy ways out, avoiding the kinky stuff, reducing complexities to one-size-fits-all. The moribund version of a language is like one of those 1920's 78 rpm records of a symphony orchestra playing, recorded acoustically rather than electrically. You get the basics, but no matter how carefully we enhance the recording with modern techniques, it's nothing like having been at the performance.

All of this is to say that, when a language dies, one of the thousands of offshoots of the first language simply grinds to a halt, after having thrived and morphed and mixed with abandon for 150,000 years.

How Do You Solve a Problem Like Revival?

Language Revival Meets the Realities of Language in Time and Space

In response to all of this, there are attempts proceeding worldwide to halt the death of minority languages, with a particularly concerted effort by many linguists in the past ten years to call worldwide attention to the problem. The effort serving as a primary inspiration is the example of Hebrew, which by the late 1800s had essentially been used only in writing and for liturgical purposes for more than two thousand years–Hebrew was an archaic-looking language encountered in weighty books, not something you had dinner in. The movement to make it the official language of Israel was so successful that today it is spoken natively by a nation of six million people. There are movements to similarly resuscitate threatened languages such as Irish Gaelic, which in many areas is taught, and taught in, in schools, with radio and television time set aside for broadcasts in the language and various activities in the language encouraged for young people. There are similar movements for Breton, Occitan, Maori, Hawaiian, and other languages. Yet these efforts, laudable as they are, face many imposing obstacles, posed in large part by the realities of how languages live in the world as we know it.

For one, as we've established, most "languages" are actually clusters of dialects. The form of a dying language taught in school is often a single, standardized variety, which can be quite different from the various dialects that constitute the "language" as it actually exists. If there is still a healthy population of people speaking the language natively and well, this "school" variety that children learn may sound rather sanitized and even imposed from without. This is a special problem in communities where the impending death of the language is a symptom of historical oppression by a surrounding power, as has been the case in Brittany with Breton. France's former policy was to discourage the use of Breton in favor of French, treating Breton as a primitive "patois" only suitable for talking to livestock. The Breton nationalist movement in response has occasionally been a violent one, and to its partisans, the alien air of "school" Breton often suffers by association with the martinet French educational tradition that has been so hostile to the language rights of Breton peoples.

We have also seen how languages mix, often when speakers are shifting from one language to another one through time. Typically, speakers leave footprints from their old language in their version of the new one (the peculiarities of Irish English come largely from Gaelic) while at the same time, during the twilight of their old language, they speak it with heavy influence from the new one. This means that, in many cases, the dying language we encounter is no longer its true self, having been tinted by the one its speakers are now dominant in.

Gros Ventre was a Native American language of Montana. When its last speakers were interviewed in the 1960s, their Gros Ventre showed evidence of remodeling on an English template. In the living language, there was no way to express the word for a body part in isolation. One could not simply say *eye*; one had to say *my eye, your eye, his eye.* The closest you could come to just *eye* was "someone's eye." Thus, to express the root *síitheh* "eye" alone would be pidgin Gros Ventre; one would have to at least say ***bi**-síitheh* "someone's eye." In true Gros Ventre, *my eye* was ***ne**síitheh.* In dying Gros Ventre, however, it was ***ne-bi**-síitheh*, where the speaker tacked the prefix for *my* onto the word meaning "someone's eye." To this speaker, more comfortable with English, in which we can say just *eye*, ***bi**-síitheh* had come to mean simply *eye* rather than "someone's eye," such that it felt natural to him to render *my eye* as ***ne**-bi-síitheh*, although to a tribal elder this would have meant the nonsensical "my someone's eye."[4]

Because it is harder for adults to learn new languages well than it is for children, when adults are forced to learn a new language quickly, the result is often various degrees of pidginization, utilizing just the bare bones of a language. This becomes a problem in revival efforts because, even when adults of a given nationality desire strongly to have their ethnic language restored to them, the mundane realities of a busy life can make it difficult to get beyond a pidgin-level competence in the language.

This is especially crucial in the language-revival case, because the languages in most immediate danger of death tend to be those spoken by previously isolated groups–for example, the peoples who were isolated enough by geography that Europeans could not transform their lands into plantation colonies in the middle of the past millennium. As we have seen, languages spoken by such groups, having had millennia to complexify without intermediation by large numbers of second-language speakers to keep the overgrowth in check, tend to be more imposingly complex than the "big dude" languages.

---

4. Which in itself sounds like a song cut from *The Music Man* on the road to New York.

To the English speaker, Spanish presents its challenges with its gender marking, conjugations, and ocasional quirks like *Me gusta el libro* instead of the "Yo gusto el libro" that would feel "normal" to us. But in general, one senses oneself as "still in Kansas"–there are plenty of similar word shapes, and how thoughts are put together is generally akin to how we do so in English. Go to languages beyond familiar ones like this and things get rockier. Someone I know who emigrated from Romania at fourteen speaks English perfectly (with a lovely hint of accent), and learning French was no problem for her. But during a stay in the Czech Republic, she ultimately decided that it was hopeless trying to pick up any Czech because, as she put it, there was simply nothing familiar: word shapes are usually unlike anything we are used to in Germanic or Romance (remember Romanian is a Romance, not Slavic, language); there is a sound or two that one essentially has to be born hearing to render properly; the nouns are declined as fiercely as the verbs are conjugated; and then there are the notorious Slavic verb pairs, where each verb takes arbitrarily different prefixes or suffixes or even changes its root according to whether actions are continuous or abrupt.

And my Romanian friend was still within Indo-European. With Native American languages, for example, one is confronted not simply with learning extremely unfamilar word shapes, but with ways of putting words together to render even the most basic of thoughts that an English speaker might barely believe humans could spontaneously resort to in running speech. *She has a big house* in Spanish translates almost word for word: *Ella tiene una casa grande.* In Czech, it is similar; typically of Slavic languages, no word for *a*, but that's not hard to get used to: *Ona má velký dům.* In general Czech "puts" things in ways that make intuitive sense to an English speaker. But then recall Cayuga's "It house-bigs her"–that's just off the scale, and remember that this is not just one wrinkle but a general way of phrasing things throughout the grammar. "Someone's eye" instead of just "eye"; having to specify just how you broke something–these are the sorts of things that confront the now English-dominant Native Americans seeking to reacquire the language of their ancestors. Here is Mohawk for *Suddenly, she heard someone give a yell from across the street:*

> *Tha'kié:ro'k iá:ken' ísi' na'oháhati iakothón:te' ónhka'k khe tontahohén:rehte'.*

Literally translated, with an attempt to make this sound the *least* needlessly "exotified," what this comes out as is "Suddenly, by what you could hear, there, it's beyond the street, the ear went to who just then made-shouted back toward her."

It's not impossible to learn a Native American language after childhood, and one can gradually wrap one's head around such ways of putting things. The language makes its own sense once one masters various general principles distinguishing such languages from ours. There are success stories of young Native Americans acquiring competence in their tribe's language from tutelage by elders, as in a long-term project directed by Leanne Hinton at the linguistics department of the University of California, Berkeley, pairing Native Americans with elders in an attempt to save as many Californian indigenous languages as possible. But learning one of these languages or an Australian Aboriginal language is hard work for someone raised in English or a related language, much harder work than picking up Spanish, and quite a job to expect of whole communities of people.

This difficulty relates to the fact that living languages are developed far beyond the strict necessities of communication and that incomplete learning guarantees that some of these baubles will be stripped away. Children learning an indigenous language in school but more comfortable in a dominant one such as English typically speak a rather simplified variety, just as do American students who learn French or Spanish in school. There is a perhaps universal tendency for elders to view youngsters as insufficiently mindful of tradition, which is heightened when the youngsters in question are assimilating to a dominant culture. This area of tension extends into children's version of a threatened language, when older fluent speakers often disparage the new version as "not real X," sometimes putting a damper on enthusiasm for the revival itself.

One of the notorious "Dammits" in Polynesian languages such as Maori and Hawaiian, for example, is the often arbitrary classification of nouns as taking either an *o* or an *a* possessive marker, and young speakers are often unsure which class a given noun belongs to. To a fluent Maori or Hawaiian speaker, this sounds like "bad" speech, just as saying *speaked* or *squoze* sounds to an English speaker. One can only imagine what schoolchildren's version of an immensely elaborated language such as Fula would sound like. The truth is that a revived language, if it "takes" and is passed on to children, will almost certainly be a considerably simplified version of the language as it was once spoken.

Finally, just as writing tends to give a language an air of "legitimacy," the converse also is true–a language that has not been traditionally written is often considered "less of a language" even by its speakers if they have been reared in a written, standardized "top twenty" language. Whereas to the scholar or social services worker, the indigenous language appears, quite properly, an exotic treasure to be cherished, to a person for whom the language is a mundane aspect of daily life, sociological realities intrude and often stamp the language as a lowly vehicle, associated with the elderly, parochialism, and a world many consider–for better or worse–a lesser option than the world of tall buildings.

Scholars have not always been immune to shades of this view. Before cultural pluralism was as overtly valued in mainstream educated discourse in America, even a linguist might describe Occitan in this fashion, this passage being from a generally masterful 1944 book on the world's languages (or the pipe-smoking Western professor of the period's conception thereof, with a Eurocentric bias focusing on standardized languages): "This *Provençal* has a flourishing culture of romantic poetry greatly influenced by Moorish culture. Its modern relatives are hayseed dialects."

If this was the best even some scholars could do until recently, then certainly lay speakers traditionally tend toward the same equation of "written" with "real." The very sound of the indigenous language immediately conveys a social context considered orthogonal to prestige, just as, no matter how Politically Correct we are and no matter what race we happen to be, we would be hard pressed to see the Declaration of Independence written in inner-city Black English as a document equal in gravity to the one Thomas Jefferson wrote. Such judgments are thoroughly arbitrary but noisomely deeply ingrained, and many communities resist efforts to revive their dying languages out of a sense that the languages are incompatible with the upward mobility they seek.

Language versus Prosperity

And this brings us to a very important matter regarding language death: why people give up their languages.

To be sure, indigenous languages have often been actively discouraged, by school policies calling for corporal punishment on any Native American student heard conversing in his home language, a practice especially common in the United States until the middle of the twentieth century, and by governmental positions declaring minority languages antithetical to national unity (witness France in the 1700s). My first office at Berkeley looked out on a courtyard called Ishi Court, named after a man who found himself the last living speaker of his native language, Yahi, after all of the other Yahis had been massacred. Many Native Americans died after similar massacres carried out by presidential administrations in the nineteenth century and depicted so placidly in the history books. Dozens of these languages expired under the watch of the Millard Fillmores and Chester Alan Arthurs by extermination or when groups were forced to live among others or to scatter, thus making it impossible to pass their languages on to enough children to keep them alive.

It is not difficult to make a case that people must not have their languages forcibly taken from them or beaten out of them. But in reality, just as often the reason groups abandon their traditional languages is ultimately a desire for resources that their native communities do not offer. Sometimes this occurs "naturally," as with the groups who now speak Maa in Kenya and Tanzania, a by no means unusual case that Western linguists would be unlikely to decry as an injustice. But more often today, it happens as a result of the pervasive effects of First World imperialism: the language of the dominant power–written, spoken by the wealthy, and broadcast constantly on radio and television–quite often comes to be associated with legitimacy, the cosmopolitan, and success. Almost inevitably, the home language is recast as, basically, *not* that–and thus antithetical to survival under the best possible conditions. This judgment is ultimately as unrelated to the stuff of the language itself as our evaluation of nonstandard dialects as "backward" is. But that's something only linguists know, for the most part, and in the meantime, many languages of Papua New Guinea, for instance, are gradually being replaced by Tok Pisin within their own largely self-subsisting villages, not through active outside imposition but because the villagers themselves have come to see this language and the access to the outside world that it offers as "cool."

Urbanization: Linguistic Slurry The trend toward urban migration that this cultural co-optation encourages is particularly lethal for language diversity. There is a short step from spending half of one's life working in a city to relocating there permanently in search of larger opportunities, especially when the degradation by large-scale logging, mining, monoculture, and other resource extractions destroys the environment a group formerly inhabited. For better or for worse, the modern geopolitical trend is toward general population intermixture in multiethnic polities. This could be termed "diversity," and indeed one potential aim might be that peoples speaking different languages will coexist within large cities, maintaining their native languages at home while using the dominant top twenty languages as utilitarian lingua francas in the realms of education, politics, and the workplace. This is the goal stated by many language revivalists, who certainly do not wish to bar indigenous peoples from the world economy.

This vision seems unobjectionable enough on its face but in reality simply could not support six thousand distinct languages. Certainly there are cities and countries where two or more languages coexist, such as English and French in Canada, Spanish and Catalan in the Catalonian region of Spain, or even more than a dozen in India. However, it is impossible, by the dictates of sheer logic, that *six thousand* languages, or anything even close, could thrive and be passed on generation after generation within the world's cities.

This is because if a city is to contain ethnic groups in a state of harmony–and presumably this is the ideal–then a phenomenon inherent to harmony is intermarriage. Love knows no boundaries, and world history eloquently demonstrates that intermarriage can only be prevented under conditions of virulent enmity between groups or, at the very least, stringent caste relations such as those in India, designating certain groups as unsuitable for intimate contact with others. The problem is that, if a couple speaking different native languages but both fluent in the dominant lingua franca marry and have children, then the children almost inevitably become more competent and comfortable in the lingua franca than in the language spoken by either parent. This is partly because the parents are more likely to speak the lingua franca to each other. Furthermore, even if the parents dutifully make sure to speak to the children only in their respective native languages, once the children are exposed to the lingua franca in school and in the world outside of the home, social evaluations kick in. Young children are exquisitely sensitive to such metrics, and quite commonly, as the child gets older, he or she begins to reject the parents' home languages in favor of the "cool" language, the one spoken by playmates, heard incessantly on television, and in general the marker of success and acceptance in the only society they have ever known. Many are the people we know who say that they spoke their parents' native language or languages when they were young but have since forgotten most of it–even when they still live with the parents and hear the language constantly. Social evaluations play a crucial role in a child's receptivity to linguistic input and orientation toward it.

Finally, even in rare cases when parents are diligent enough to maintain their child's fluency in the home language or languages, when that child himself marries, the chances of his marrying someone who speaks the same language (or certainly languages) are slim, and hence the chances that *their* children will speak a language they hear only occasionally from one parent and otherwise only when grandparents visit is nil.

The Grass Is Always Greener: The Mundane Realities of the "Exotic" What this means is that a world where all six thousand of today's languages thrived would be, properly speaking, one where a great many peoples remained rooted to isolated hunter-gatherer, pastoral, or small-scale cultivational existences, untouched by the First World. There is perhaps a certain romance in that idea: we are trained to emphasize the downsides of our First World existences, which are certainly many, such that a picture of a globe peopled by smallish groups living on the land may seem "the way things should be." Yet for all of the pernicious injustices and psychological dislocation inherent to Western life, there is perhaps a danger in romanticizing Third World cultures as well. The rain forest–dwelling Amazonians whose cultures and languages are dying at an alarming rate are, after all, also societies where life expectancy is often brief, diseases easily cured in the West are often rampant and lethal, there is a high infant mortality rate, and the treatment of women would be unthinkable to anyone reared in a "modern" society, especially since the 1970s. It is not accidental that it is often women in rural and indigenous societies who are in the vanguard of opting out of the native language in favor of the linguistic key to success in the surrounding culture, where women have more freedom of choice about child rearing and more control over their relationships with men.

In the conclusion of a much-discussed article in academic linguistics' signature journal *Language*, the eminent linguist Peter Ladefoged described a Dahalo-speaking father who was proud that his son now spoke only Swahili, because this was an index of his having moved beyond the confines of village life into material success beyond. "Who am I to say that he was wrong?" writes Ladefoged. Certainly we cannot prefer that the son opt for poverty; he was most likely moving away from a context in which few Westerners, language revivalists or not, could even conceive of living if other options were available. This last point is crucial: even if the Dahalo speaker was seduced by attractions of city life and the cash economy that in our eyes are of superficial value, we put ourselves in a tenuous position when we argue that the son should resist the very life style that none of us, downsides fully acknowledged, would even consider giving up.

The possible objection that it would be preferable for his son to move to a city but be bilingual in Dahalo and Swahili would most likely be a stopgap solution: dozens of languages are spoken in Tanzania, and the chances that the son will marry a fellow Dahalo speaker in the city are slim. In response, Nancy Dorian, who spearheaded the study of language death with seminal work on the demise of Scottish Gaelic, answered Ladefoged by noting that the generations sired by the one that let its language go often come to resent their parents for not passing on such a precious inheritance. But the sad question is whether their having tried to pass the language on would have been effective in a context where those very children would have been lapping up the dominant language as eagerly as all children do–and if they had, would the children have been able to pass the language on to *their* children?

Practical Solutions

Cognizant of these problems and paradoxes, Daniel Nettle and Suzanne Romaine argue in *Vanishing Voices,* the most deeply thought of the various book-length treatments of the language-death matter, that any realistic worldwide language-revival effort must take place within a general initiative allowing indigenous groups to continue living on their lands within their own cultures. Nettle and Romaine view language death as a symptom of the larger process of the rape of the world's landscapes and the destruction of the cultures that once thrived within them, driven by the insatiable capitalist thirst for natural resources and the often brutally centralized control necessary to ensure its continual slaking. By no means so utopian as to require that native groups not acquire top twenty languages in order to participate to some extent in the world economy, Nettle and Romaine propose that such groups be ushered into a diglossic use of dominant languages and their native ones. Their point is that only if such groups are encouraged and allowed to stay in their traditional settings will such diglossia not be a mere stopgap along the way to the abandonment of the "low" language forever in favor of the "high."

Nettle and Romaine's message is as depressing as it is sensible, because at its heart is the belief that the preservation of any significant number of the world's languages will require a significant transformation in the global economy, which is driven largely by governments for whom such notions as cultural diversity have been anathema at worst and of low priority at best. The sad fact is that Western scholars' earnest musings on the value of linguistic diversity are ultimately a luxury of the prosperity created by the very destructive policies at the heart of the extinctions in question. It is not accidental, for example, that to date almost all of the seminal books and anthologies on language death have been published by Cambridge University Press, an entity representing and founded on an institution made possible only through the wealth generated by what was once one of the world's most nakedly imperialist, exploitative powers.

Developing countries, constrained by limited budgets, pressing poverty, and poor educational systems, and too frequently run by despotic dictatorships as little concerned with minority rights as the monarchies that created today's First World countries, generally only pay lip service to European calls that they preserve their lands and indigenous cultures. After all, the very European countries urging "multiculturalism" on, say, an Indonesia only developed their own broad-horizoned intelligentsia on the basis of resources derived from deforesting and polluting their own countries, as well as others, and often exterminating other cultures in the process.

This is not to say that we coddled Western intellectuals are wrong in our exhortations. It is clear to many that cultural relativism has its limits, and I believe that we can assert that the preservation of environments and indigenous cultures is a desirable pathway for humankind without censuring ourselves for imposing "ethnocentric" conceptions. I for one can quite confidently reject the notion that the erasure of the entire Amazonian rain forest be treated as a legitimate expression of a "different culture." I would love to see Nettle and Romaine's articulate exhortation and its general frame of reference serve as the foundation of increased efforts to prevent most of the world's peoples from being subsumed into a slurry of multiethnic urban misery and exploitation voiced in just a couple of dozen big fat languages.

What Will Happen to the First Language's Children?

Yet it is clear in view of modern realities that a great many languages now technically alive will not be saved. A sober yet progressive assessment of the situation might be that today's endangered languages constitute three main sets having potential viability.

Many Will Either Survive or Become Thriving "Taught" Languages

In the relatively successful language revivals of Irish, Breton, Maori, Welsh, and Hawaiian, large numbers of children are learning the languages in school, the media have joined the effort, and there are increasing amounts of printed materials available in the languages. However, it is also true that these languages remain very much the second languages of most of the learners, not much spoken at home or in casual situations. Whether these languages will survive as natively spoken ones is at this writing essentially a question mark.

Hebrew was indeed revived from the page at the founding of Israel–but the fact that today this case remains the only one commonly referred to as a success story signifies that it was unusual. The revival of Hebrew was favored by its occurrence within a new country where the language was explicitly designated as the intended official one, with the government expressly committed to the effort rather than setting aside occasional funds for the use of Hebrew "alongside" another language. Furthermore, the original immigrants to Israel spoke various languages and thus there was a motivation for a new language to express the new national identity, in contrast with Welsh, Irish, Hawaiian, or Maori people who, for better or for worse, can only adopt the indigenous language as an "add-on," English having long been their primary language. Finally, the adoption of Hebrew was assisted by its link to a religious tradition, virtually a covenant: even though Hebrew was declared the official language of Israel from on high, the success of the movement was determined by a powerful sentiment within the families themselves that the use of this language was critical to the establishment of a Jewish state. It has been said that the revival would not have succeeded without this crucial element–a sense of learning Hebrew as an imperative of, again, one's very soul, not just as a kind of party trick or "local custom."

Conditions in Ireland, Wales, New Zealand, or Hawaii only approximate the spiritual ones that reanimated Hebrew. The indigenous languages are not connected to religions still alive and deeply felt by most, nor are the revival efforts taking place among a people committed so starkly and universally to cultural sovereignty as to relocate to a brand-new nation or be allowed by sociohistorical serendipity to found one. Yet many of the people learning these languages feel that they are not fully expressing their souls without speaking the indigenous language. This is a hopeful sign. In all of these places, increasing numbers of homes are passing the language on to their children as a first one.

Yet even if these success stories remain too scattered to revive the languages generally, their situation is not quite as hopeless as is claimed by various commentators who have declared the Irish revival movement a failure because of the unlikelihood that significant numbers of families will pass the language on to children as a mother tongue. It is a central tenet of the language-revival movement that a language is only truly alive when it is regularly passed on to children, but this is not necessarily true. More properly, throughout human history thus far, this has been the case. Yet it is conceivable that languages such as Irish, Welsh, Maori, and Hawaiian could be passed on as second languages, taught in school and spoken nonnatively but proficiently, *in perpetuo.* Under such conditions, the languages could persist as cultural indicators, the very learning of the language in school itself constituting a hallmark of cultural identify. As such, the population would surely speak the language with varying degrees of proficiency, some excellently, others only controlling the basics (as do many Americans in California who "speak Spanish" as the result of a few years of classes in school followed by constant exposure to the language from the large Latino population), and many people falling somewhere between these poles.

This is, after all, the case with many lingua francas in Third World countries, with more speakers having learned the languages as teenagers or later than having learned them natively; Swahili has long been an example. Many languages born as pidgins have been spoken as nonnative languages for centuries, learned mostly by men in work contexts and quickly expanding through constant use of this kind into creoles, suitable for precise and modulated expression. It is perhaps something of a Western conceit to suppose that a language is not "a language" unless it is spoken from the cradle. This requirement, after all, would imply that clergy speaking Latin or Sanskrit are not really "speaking the language," because they did not learn the language as infants, a claim that would ultimately seem to be rather arbitrary. Similarly, Africans typically "speak" many languages that were not spoken to them until adolescence or later; even if a speaker's version of a language constitutes only, say, seventy-five percent of what a natively transmitted version consists of, it is unclear that this African "does not speak" the language.

For better or for worse, the cultural conditions are present to preserve, for example, Irish within what could be considered a domain for minority languages commensurate with new world conditions: a living *taught* language. The invention of writing, which has threatened minority languages in tending to anoint the dominant languages chosen to write in as "legitimate," can ironically be of assistance here, in allowing the transcription and dissemination of language-teaching materials.

Many Will Likely Survive Only as Living "Taught" Languages

Then there are the languages concentrated in tropical regions that are threatened by the encroachments of global capitalism–the

Dahalos of the world, such as the Ugong language that Thai is edging away, and the more than eight hundred fabulously complex and variegated languages of Papua New Guinea. In these cases, the glass-half-full perspective suggests hope that national governments can be persuaded to assist in preserving the cultures speaking these languages, because only this will allow them to continue to be spoken natively. On the other hand, a constructive response to the glass-half-empty perspective, conceding that brutal realities make it likely that such attempts will not be able to save anything approaching all of these languages, would be to adopt the "taught language" perspective, providing for a time when descendants of today's native speakers will at least be able to acquire some proficiency in their languages through schooling, to the extent that the descendants remain a coherent enough entity to ensure a suitable demand.

A Historically Unprecedented Number Will Die

On the other hand, almost all of the indigenous languages of North America and Australia would appear to be lost forever as living languages. All but a handful are spoken only by the very elderly, as foreign and imposing to many of their English-dominant children and grandchildren as they are to us. In most cases, surviving descendants of a given group are too few and too geographically scattered for there to be significant demand for revival of any kind in the future.

A Really Good Chinese Restaurant in San Francisco

In general it would appear that the linguistic landscape of the future will be a less diverse and somewhat blander one than has existed until now. Many of the languages that survive as natively spoken will be mostly geopolitically dominant ones, and such languages, by the very nature of having through the ages been learned by large numbers of adults and as often as not used as secondary rather than primary languages, are often somewhat "streamlined" in regard to grammatical elaborations. This means that a certain "vanilla" quotient will be overrepresented among the surviving languages–recall that Swahili is somewhat watered down in complexity as Bantu languages go; it has even been argued that the Romance languages, representing Latin learned as a second language by subjugated populations, are slightly "pidginized" in comparison with other Indo-European languages. Note also that Wolof, in becoming the lingua franca of Senegal, is probably on the way to seeing its array of noun class markers severely reduced as a "price" to pay for its new broadened sway–power corrupts! Meanwhile, a substantial number of minority languages will persist in use as "taught" languages–but then in this guise these languages will be somewhat less elaborated than they were when spoken natively.

One might analogize the linguistic landscape of the future to a world where the dazzling variety and subtlety of native Chinese cuisines, the product of thousands of years of accumulated skills, evolutions, branchings, and mixture, are represented only by Chinese food as available in the United States. Certainly, a great deal of excellent Chinese food is available here, but not in the protean richness available in China, and a great deal of what Americans are accustomed to eating as "Chinese" food is actually better described as Chinese ingredients adapted to a beef-stew palate. Yet just as this is surely better than nothing (there was no won ton soup, sushi, coconut milk soup, or even spaghetti and meatballs served on the *Titanic* in 1912), the admittedly blanched language palate that even our most dedicated language-revival efforts will most likely leave behind is certainly better than what would remain if we did nothing.

The Task Ahead and Why It Must Be Done

It is therefore urgent that we record as many languages as possible before they no longer exist so that, even if they are not actively spoken anymore, we have their essences preserved for posterity for the benefit of descendants of speakers who want to make contact with their heritage by learning some of the language; for research; and for sheer wonder.

It is here that linguists, the people most qualified to carry out this task, will be crucial, but only if there is a fundamental recasting of current attitudes in the discipline. People often suppose that linguists are either professional polyglots or arbiters of "proper grammar." Neither is the case; in fact, precisely what most linguists are engaged in would surprise many people by virtue of the extremely specific nature of the enterprise, focused on a particular issue barely perceptible at all to the layperson.

The linguistics discipline as it is today configured is centered on identifying through elegant induction the precise structure of our innate neurological endowment for language, sparked by a paradigm founded by Noam Chomsky in the late 1950s. There are many other branches of linguistics and a great many linguists with no serious interest in the Chomskyan approach. However, the paradigm looms over the field with a sociological "capital" analogous to the domination of the music composition field decades ago by atonalists despite their never having been a numerical majority.

One's basic training focuses on the Chomskyan framework, and there is a tacit but powerful sense in the field that this subarea is not only the "sexiest," but also the most intellectually substantial. For example, there are some departments where students are trained in nothing but the Chomskyan paradigm, but none where students are grounded entirely in any other subfield–the other subfields are ultimately regarded as "other," the icing rather than the cake. Regardless of the caliber of his work in another subfield, the linguist who does not display at least token interest in the Chomskyan endeavor is not considered "a *linguist* linguist" in the back of the minds of a great many in the field, and the most general respect is accorded the linguist in an "icing" subfield who is invested in showing the implications of his work for the latest developments in the Chomsky bailiwick. For example, it is safe to say that to most modern linguists in America the phenomena I have covered in this book are perhaps "interesting" in a passing way, but generally not considered "real linguistics."

To be sure, Chomskyan linguistics is a thoroughly fascinating investigation. Steven Pinker's book *The Language Instinct* should in my opinion, along with Jared Diamond's *Guns, Germs, and Steel* (a rare example of a book that tells us what we want to hear and is empirically correct in the bargain), be required reading for all thinking people. It is not for nothing that the Chomskyan paradigm took our field by storm to such an extent in the 1960s and has obsessed so many fine minds since. Properly, however, illuminating the possibility that we possess a neural mechanism calibrated to produce basic sentences is but one of dozens of ways that one might study the multifarious thing known as human language. In our moment, as linguist R. M. W. Dixon eloquently calls for in his book *The Rise and Fall of Languages*, linguists should be trained to go out and document at least one dying language before it disappears forever from the earth–I myself will be embarking on such work as soon as I finish this book. This is particularly appropriate given that the study of such a language inestimably enriches the study of the possibility of an innate language competence, often furnishing a career's worth of relevant data. (Notice my sense of obligation to say that, so powerful is the sense of "Chomsky–smart/other–also-ran" in modern academic linguistics in the United States.)

It is often said that we must preserve the world's languages because each one reflects a particular culture. Although this is true in itself, I have always felt that to elevate this as a guiding motivation for preserving languages is based on an oversimplified conception of the relationship between language and culture. It is true that when a group loses its language millennia of accumulated knowledge regarding the medicinal properties of plants, the subtleties of managing crops, the life cycles of fishes, and other phenomena are lost. However, in the strict sense, the linkage of language revival with cultures seems to imply that once researchers recorded the cultural aspects of language for posterity, then it would no longer be important whether or not the language as a whole continued to be spoken.

And in any case, as I have noted previously, most of a given language has evolved less on the basis of culture than through the structured randomness of an evolution bounded only by human physiognomy and cognitive requirements. All but a few pages of any written grammar of a language is taken up with elaborate rules, lists, and exceptions that no more reveal anything specific to the culture that uses them than a pattern of spilled milk reveals anything specific about the bottle it came from.

Linguists are quite aware of this, and in fact most linguists' scholarship on languages has little to do with charting links between grammars and cultures. It is safe to say that most, although not all, linguists largely cherish languages because of the sheer marvel of their various architectures, elegantly combining structure and chaos in six thousand different ways. I surmise that the emphasis on culture among linguists active in the language-revival movement stems from a sense that the purely linguistic wonder of human speech is less accessible to the general public than arguments founded on more easily perceived concepts such as culture.

Yet throughout this book I have hoped to usher the reader into the very awareness animating linguists that human speech is a truly wondrous thing in itself. In this vein, it pays to note that the Dahalo language that Peter Ladefoged referred to is the one with clicks spoken far from the territory where the other click languages are spoken–the language the farmer thinks of as a sign of backwardness is, with all due respect to his justifiable relationship to his immediate circumstances, a language with a wondrous sound system. A great many of the Native American languages dying before our eyes were so complex that children were not fully competent in them until they were ten years old. It is truly sad that world history cannot allow all of these languages to continue to be spoken, transform themselves into new ones, overgrow, and mix with one another. But at the very least we can make sure that as many of them as possible are written down as thoroughly as possible before their demise as living systems and that at least a healthy number of lucky ones can be passed along as secondary but essential languages across generations.

Let's take a look at one last descendant of the world's first language. Because prefixes and suffixes generally evolve in a language from what begin as full, separate words, the first language can be assumed to have had no prefixes or suffixes at all (or tones or a great many other complications of a grammar that only arise through gradual reinterpretations of material). Yet 150,000 years later, gradual evolution produced a remarkable array of prefixes in the Central Pomo language of California. English speakers associate prefixes with relatively basic meanings such as repetition (*re-*) and opposition (*un-, in-, mis-*). But in Central Pomo, prefixes carry much more robust and specific meanings:

ba-	orally
s-	by sucking
š-	with a handle
ča-	by slicing
čʰ-	pertaining to vegetative growth
da-	by pushing with the palm
h-	by poking
m-	with heat
qa-	by biting
ša-	by shaking
'-	by fine hand action, such as using the fingers

The root *yól* means "to mix." Each of its combinations with these eleven prefixes yields a particularly useful word:

ba_yól_ — to insert words suddenly while humming; that is, mix orally

s_yól_ — to wash down cookies or doughnuts with coffee; that is, to mix by sucking

š_yól_ — to stir with a spoon; that is, to mix with a "handle"

ča_yól_ — to chop up several things together, such as celery and onions for stew

č^{h}_yól_ — to plant things close together

da_yól_ — to fold in dry ingredients while baking

h_yól_ — to add salt or pepper (I guess they "poke" it in)

m_yól_ — to throw various ingredients into a pot; that is, to mix by heating

qa_yól_ — to eat several things together, such as meat and potatoes; that is, to mix by biting

ša_yól_ — to sift dry ingredients

'yól — to throw ingredients into a bowl with the fingers

Of course, the prefixes create new words with each verb; *'ól*, with a glottal stop as its first consonant, means "to summon," and here are some of its prefixed versions:

ba_'ól_ — to call; that is, to summon orally

š_'ól_ — to set a fishing line; that is, to summon by manipulating a handle

č^{h}_'ól_ — to comb hair; that is, to summon vegetative growth (by analogy with the flowing motion of some vegetation)

da_'ól_ — to dig for; that is, to summon by pushing with the palm

h_'ól_ — to probe for a creature with a stick; that is, to summon by poking

It is this kind of thing, then, that we are losing when languages die–the last known fluent speakers of Central Pomo have died since these data were collected. Just as we would be inestimably poorer to be denied the opportunity to see giraffes, roses, bombardier beetles, tulips, and little black house cats with white spots on their chests that sit on our laps as we write, we lose one of the true wonders of the world every time one of these glorious variations on a theme set by the first language slips away unrecorded for posterity. We will never encounter a stegosaur, but we can be thankful that fossils allow us to know what it was like. In the same way, if we cannot enjoy all six thousand of the world's languages alive for much longer, let us at least make sure to afford them high-quality preservation.

In the Central Pomo case, certainly the loss of the language entailed the loss of a vehicle of cultural expression. But surely all of us value sucking, poking, and shaking as much as the Central Pomo speakers did: it's just that our languages chose not to index such things with prefixes. Most likely the reader's native language chose instead to genuflect to marking each noun as definite or indefinite. Both the Central Pomo prefixes and the European languages' articles are fascinating in their own right as alternate methods of packaging information in order to talk about this thing called living, and both are only the tip of the iceberg in regard to the endless ways in which humans can express themselves in speech.

Each variant of the first language is festooned with gloriously random remnants of things caught in the cracks in the course of transformations long forgotten, and most of them exist in an array of subvariants on the theme related to one another rather like Barbara Cartland's hundreds of romance novelettes. All carry mementoes of past liaisons with other dialects of other languages; some of them once rose from the ashes; most of them developed as far beyond the call of duty as the Cathedral of Notre Dame. A select few even sit swathed in a Dorian Gray complex as a by-product of the invention of the printing press. The world's riffs on basic materials that emerged in East Africa around 148,000 B.C. represent six thousand ways of being human.

Epilogue: "Extra, Extra! The Language of Adam and Eve!"

In light of the past seven chapters, I feel obliged to address a question many readers may have had throughout the book. I have only referred to the world's first language in the abstract. Yet readers of the *New York Times,* the *Atlantic Monthly,* and other publications might ask, "Haven't words from the world's first language been reconstructed?"

Indeed, the eminent linguist Joseph Greenberg, followed by Merritt Ruhlen in conjunction with John Bengtson, proposed a list of twenty-seven words they consider likely to have been used in the ancestor to all of the world's languages. They reconstructed these words by using the general approach we saw for arriving at the word *snusos* for "sister-in-law" in Proto-Indo-European: deducing backward from words for these concepts in the languages of the world. They propose, for example, that the original word for *finger* was approximately *tik.* They include in their data words not only for *finger* but for related concepts such as *hand,* to *point,* and the number *one* (symbolized with one finger) quite reasonably, given the centrality of semantic change in language transformation (remember how *silly* started out meaning *blessed*). As one travels around the world, one does find plenty of words for these concepts that approximate *tik*: the Dinka of Sudan have *tok* for "one"; the reconstructed Proto-Indo-European word for "to point" is *deik,* which comes down to Latin as the *dig-* in *digitus* "finger"; Tibetans have *(g-)tśig* for "one"; and there are similar forms for "one," "finger," or "hand" in many of the several hundred languages spoken by Indians in the Americas: *tok, zek, tinki, dooki,* t^s*ikitik,* etc. I have given only a fraction of the examples cited by Greenberg, Bengtson, and Ruhlen, and the last two bring a similar range of data to bear for another twenty-six words.

Despite their frequent and sympathetic coverage in the media, these "Proto-World" reconstructions are not considered valid in the linguistics field in general. Greenberg and particularly Ruhlen, the primary spokesman for the hypothesis, portray this as a classic example of blinkered, tradition-bound bean counters unable to see the value in fresh, wide-lens ideas. The point of contention is method of analysis. Traditionally, language-change specialists methodically and exactingly deduce protoforms by the kind of close reasoning we saw with Proto-Indo-European *snusos.* Indo-European is only one of many language families in the world, and some of the other families' protolanguages have been reconstructed in the same fashion as Proto-Indo-European. However, both traditional historical linguists and the "language of Adam and Eve" contingent agree that it is impossible to deduce what the first language was like by this traditional method. This is for the simple reason that the family protolanguages that we can reconstruct only go back several thousand years in time, whereas the first language originated several tens of thousands of years before that. Not only have most families' protolanguages not been reconstructed in any detail, but any reconstructed protolanguage is a good guess at best. And on top of this, untold thousands of languages and certainly at least dozens if not hundreds of whole language families have died out through the millennia, depriving us of the data with which to deduce what ancestral forms farther back might have been.

Greenberg and Ruhlen object that this kind of deductive reconstruction is only one way at arriving at what they call "Proto-World," and that uncanny correspondences across all of today's languages are evidence of the Ur-language just as valid as step-by-step deductive reconstruction. As Ruhlen puts it:

> Were a biologist to demand a complete reconstruction of Proto-Mammal, together with a complete explanation of how this creature evolved into every living mammal, before he would accept the fact that human beings are related to cats and bats, he would not be

> taken seriously. Yet it is just this kind of linguistic nonsense that has been taught in universities by Indo-Europeanists for so long that most linguists are unaware of its mythological nature.

It's a grand old story–unimaginative, indoctrinated "terrible turtles" (in Dmitri Nabokov's meaning) too caught up in their reputations to see the forest for the trees and open up to ideas that go against the grain. We think of Copernicus, Galileo, Darwin, and Alfred Wegener, whose theory of continental drift was scornfully ridiculed by geologists until magnetic patterns in rocks and correspondences in fossil finds proved him correct. Language-change specialists think of founding linguist Ferdinand de Saussure, who from quirks in living Indo-European languages deduced that Proto-Indo-European must have included a set of throaty consonants, even though these are not found in any of its living descendants. His hostilely skeptical colleagues wouldn't let poor Ferdinand join in any more reindeer games until the unearthing of documents in an extinct, hitherto unknown Indo-European language, Hittite, revealed those very consonants in black and white.

Seeing unknown linguists' snippy dismissals of Greenberg juxtaposed with the sexiness of the possibility of knowing that Adam and Eve's word for *finger* was *tik*, *I* was *n*, and their way of asking "What?" was *Ma?* (fortuitously just the word for *what* in the Hebrew that they were first depicted as speaking) or reading Ruhlen's crisply written and nobly indignant précis of the Proto-World hypothesis for the general public *(The Origin of Language)*, one *wants* to go with Greenberg and Ruhlen. Apostasy is attractive to the outside observer. I personally am definitely a lumper rather than a splitter and have an inclination toward giving heretical theories as receptive a hearing as possible. My work in linguistics (and beyond) has entailed a fair amount of tilting at windmills.

But sadly, the Proto-World reconstructions, in the end, are hopelessly untenable, even from a perspective devoid of any "Harrumph!" The sympathetic coverage that this work has received from the media is due to their being understandably titillated by the idea, just as understandably inclined toward the "Visionary mavericks battle the hidebound establishment" hook, and not in a position to review the facts. But these articles innocently distort the mundane and irrefutable truth that, short of either a time machine or unprecedented leaps of refinement in chaos and emergence theory, we will never know any words of Proto-World.

The main problem is a very simple one: 150,000 years of language transformation by thousands of offshoots of Proto-World are certain to have hopelessly obscured any sign of what any word in that original language would have been. Ruhlen's analogy to the mammalian evolution case is deft but infelicitous. It's not that linguists are closed to the idea that all languages began with one and are all thus distantly related. It is safe to say that most linguists believe this. Their objection is that we cannot specifically *reconstruct* anything in that first language. We have seen, for example, how, in roughly a millennium, Latin turned over so thoroughly that it became a new language entirely: "FEH-mee-nah" became "FAHM," *de de intus* [deh deh IN-toos] became *dans* [DAW^{ng}] for "in," obscure words rose to oust common ones to express basic concepts, such as *parabulāre* replacing *loqui* for "to talk." We have seen how *hussy* began as *housewife* and *every* as *ever each*. Here's the rub: we could never know of these things if Latin, French, and English had not happened to be written for so long. For example, there is nothing about *dans* that would clue the linguist into the fact that it began as three words, much less just what those three words were.

With shallow time depths, the linguist can and does make informed guesses about the earlier states of given words. By comparing cognate words in all of the Romance languages, for example, linguists can often reconstruct the original Latin word even when that word is not actually attested in Latin texts. But then look at what happens through several millennia. Theories about when Proto-Indo-European was brought to Europe vary, ranging from 6,000 to 9,000 years ago. And even in this sliver of the 150,000-year history of human language, recall that the word for *sister-in-law* evolved from *snusos* to *nu* in Armenian and no descendant of the *snusos* root even exists any longer in modern English. Or take a simple concept such as *bread* in various Indo-European languages: *pan* in Spanish, *xleb* [KHLYEP] in Russian, *psomi'* in Greek, with many of these languages as well as others in the family having replaced the original root with another one, *parabulāre/parler*-style. Languages morph continually, always have, and do so even faster in preliterate societies, where written forms are not enshrined as "standard" and retard change–recall the Ngan'gitymerri example in Chapter 5. And of course, rapid change was typical in all languages until the invention of writing just 6,000 years ago, at which point human languages had already existed for about 145,000 years!

Cheyenne of Montana and Oklahoma is one of those unwritten (until linguists got to it) languages, one of several of the Native American Algonquian family, whose protolanguage has been reconstructed. Cheyenne provides a useful example of the sheer thoroughness of language transformation. The word for "winter" in Proto-Algonquian was *peponwi* [peh-PONE-wee]. However, this word really went through its paces on the way to Cheyenne, all results of ordinary sound changes piling on top of one another.

First, the last syllable was dropped just like the *-e* of *name*, resulting in *pepon*. Then the middle *p* dropped out (much as the *p* of Latin *super* dropped out when it became the prefix *sur-* in French words like *surabondance* "superabundance"), and the first *p* fell off (consonants can do this, like the *h* at the beginning of words such as *Hereford* for which Professor Higgins was so annoyed at Eliza for letting go). The result now was just *pepon* without the *p*'s: *eon*, pronounced "ay-OWN." Then the vowels started shifting as they did to turn Latin "KAH-nem" into French's "SHYEH^{ng}" and Old English "NAH-muh" into today's "NEIGHM." The particular shifts in Cheyenne turned "ay-OWN" into "ah-EEN": *ain*. Then the final *-n* dropped off–typical stuff, but now all we have is "ah-EE": *ai*. Then came a stranger change: the final *-ee* cloned itself and developed a glottal stop to separate the two twins, the result being *ai'i* [ah-EE-'ee], where the ' symbol is the glottal stop; put the same catch in the throat between the *ee*'s that you do between the two syllables of *uh-oh*. Then by the "magnet" effect we saw with Proto-West Germanic *gosi* becoming *gøsi*, where the *o* became distorted through speakers' anticipating pronouncing an *i* in the next syllable, *ai'i*'s first *i* evolved into an *a* because it came right after one: ***aa'i*** [ah-AH-'ee]. But you know what tends to happen to final sounds without an accent, and so eventually today's word is just *aa'*, pronounced roughly "ah-AH'," with the catch in the throat still hanging on the end. Now remember where we started: *peponwi*! From *peponwi* to *aa'* in only about 1,500 years.

I recall my father winning a Monopoly game in a half hour through a careful series of individually ordinary strategies that concluded in his killing me with a rather downmarket hotel. I couldn't quite accept that this unorthodox sequence of actions could have foiled my usual approaches, but his explanation made it as clear as day that he hadn't done anything except follow the rules. In the same way, ordinary language change turns *peponwi*'s into *aa'*'s all the time just step by step, century by century, millennium by millennium:

p e p o n w i	Buy Connecticut Avenue to match Oriental.
p e p o n	Trade a random green for Vermont Avenue with John.
e o n	Now I have all the light blues; John only has two greens.
a i n	Build hotels on them; John thinks they're small potatoes.
a i	John gets the third green.
a i ' i	John overspends putting hotels on them, expecting quick payback.
a a ' i	But my hotels keep nailing him as he passes "Go."
a a '	John lands on Connecticut twice in a row and bites the dust.

In reference to Proto-World, then, our question must be: Given how thoroughly all languages are constantly transforming themselves, why would we expect even a single form from the original language to survive? If we plug in a lava lamp tonight, how likely is it that a week later even a single thread of that goop is going to be in the same place it was when we turned the lamp on? That not a single form of Proto-World would survive is precisely what we *would* expect.

Now, Greenberg and Ruhlen might object that, contrary to this expectation, apparently some forms such as *tik have* survived and that this evidence must be given its weight even without an explanation for their failure to change, just as finding a Honda Civic on Mars would require accepting that humans had been there whether or not we knew how. But the problem is that *tik* and the others are not Hondas on Mars. They are more like "horsies" seen in cloud formations on separate occasions: if a cloud we see looks like a horsie on Tuesday and then another one does on Friday, we assume that different clouds happened to take that approximate shape on separate days by chance, not that one horsie cloud somehow rested immobile while the other clouds kept changing shape and blowing away.

That is, we can only conclude that similarities in shape between words for the same concept are evidence of an original template if the similarities cannot be due to chance. Unfortunately–truly unfortunately, because it would be so much fun if things were otherwise–chance would quite readily lead vastly separate transformations of the Ur-language to coincide here and there in the shape of words for the same concept. There are only so many sounds the human mouth can make, and only so many possible combinations of those sounds, especially when we factor in that words get only so long, that only so many consonants can be clustered together and be pronounceable (Russian's *vsgljad* is about the limit), that only so many vowels can be juxtaposed (no language has words like *eoiaoiueraaaaiuuiao*), that the number of vowel sounds a language can have is relatively limited, and so on. Because of this and the fact that there are several thousand languages, all of them the product of several tens of thousands of years of evolution, it's not that words in some of them *might* coincide but that we would *expect* them to. In other words, for no words to coincide would be unusual and would in itself constitute a mystery to be investigated.

For example, the words in Thai for *fire, die,* and *rim* are *faj, taaj,* and *rim,* just by accident! Long lists have been composed of correspondences like this between hopelessly disparate languages; it can be almost funny. According to the Proto-World advocates' modus operandi–allowance for stark differences in word shape and a permissive position on what constitutes related meaning–English and Japanese could be shown to have a historical relationship according to these words I have always noticed:

JAPANESE	MEANING	ENGLISH
mō	more	*more*
sō	like that	*so* (as in *just so*)
sagaru	hang down	*sag*
nai	not	*not*
namae	name	*name*
mono	thing (a single entity)	*mono-* "one"
miru	see	*mirror* (which one sees in)
taberu	eat	*table* (where one eats)
atsui [ott-SOO-ee]	hot	*hot*
hito	man	*he*
yo	emphatic particle	*Yo!*
kuu	"feed your face"	*chew*
inki	dark spirited, glum	*inky* (dark)
o	honorific prefix	*O* ("O, mighty Isis")[1]

And these are just the ones I can think of in passing; there are surely more. Linguist William Poser has even formally demonstrated that correspondences like these even across several languages of the world are statistically predictable, given the constraints on sounds and how they can form words.

It is often said that chimpanzees kept alive in a room for a million years with a typewriter would at some point bang out *Hamlet* by sheer accident (ideally typing *thrift* instead of *husbandry*). More to the point here, imagine six thousand chimpanzees kept alive in an airplane hangar for 150,000 years, each chained to its own typewriter with a continuous roll of paper fed into it (all due apologies to animal rights activists, but then I almost feel as if I'm describing myself). Not only would all of them frequently happen to type the same little sequences of letters, but many of them would be found to have done so at roughly similar points on the rolls of paper.[2] To wit: linguists have an explanation for the *accidental development* of scattered correspondences between words in the world's languages as each language transformed eternally on its merry way: chimps–

---

1. Does anyone remember this from the live-action Saturday-morning show from the '70s? The lead actress, Joanna Cameron, helped to awaken the kind of stirrings in me that eventually change a ten-year-old's life forever.

2. Chimpanzees have always made me extremely uncomfortable.

I mean, chance. The Proto-World defenders fail to explain why any one word from the first language, much less several, would have simply sat and *resisted change* in certain languages since the origin of language more than 150,000 years ago.

To be as fair as possible, Greenberg and Ruhlen base their case not only on commonalities scattered across all of the world's languages, but also on commonalities among twelve reconstructed protolanguages, representing most of the language families in the world. These protolanguages are naturally a much smaller group than the six thousand living ones, and therefore, theoretically, coincidences between them carry more weight in a theory of origin. But the problems remain even here: the data they appeal to in citing these protolanguages are extremely shaky. All of the details would tax any but the few hundred specialists in language change worldwide, but two quick samples illustrate the nature of this methodological problem.

Not all European languages are of the Indo-European family; Finnish, Hungarian, Estonian, and Lappish belong to the Uralic family, which has several other more obscure members stretching eastward into the former Soviet Union. Bengtson and Ruhlen cite *ik, odik,* or *ɤtik* (pronounced "oo-teek," with roughly the *oo* of *foot*) as proposed Proto-Uralic forms. But the reconstruction of Proto-Uralic is only in its infancy, and this form was actually reconstructed for only one *subbranch* of Uralic and is thus no more "Proto-Uralic" than the first horse was the "Proto-Mammal." Only by using data from the other subfamilies in the group could we reconstruct the forms of "Proto-Uralic." But crucially: if the protoforms of the other subfamilies were, say, *arg*, *pylk*, and *zor*, then the Proto-Uralic form would have to be reconstructed as something that both they *and odik* or *ɤtik* could plausibly have descended from. This could no more plausibly have been simply *odik* and *ɤtik* unmodified than the common ancestor of crocodiles and birds could have been just some bird–you need something intermediate between the two branches. This same problem haunts other purported family protoforms in the Proto-World work.

Bengtson and Ruhlen present *nto'* as the protoform for a small group of languages spoken by isolated groups in Southeast Asia including Hmong, called Miao-Yao (a name particularly cute, given that I am writing this with my cat on my lap, as usual). But this form was not arrived at through the careful sifting for a century and a half by hundreds of scholars devoting their careers to the enterprise as Proto-Indo-European forms have been. The Proto-Miao-Yao form is only given as a tentative one by one scholar. Certainly, we cannot require that any protoform recruited in the service of a larger conclusion be the product of 150 years of collective research begun by continental scholars with mustaches. But *nto'* is ventured in preliminary fashion on the basis of no fewer than eight possible Proto-Miao-Yao forms suggested, all of them in turn deliberately left full of possible alternatives to be resolved by further work, which is what the brackets and parentheses and slashes indicate: *tu(ń)źuk, t/l/uńźuk, (n)tu(ń)źuk, tu(ź)i(k), [tu](ń)źi[q], tu(ź)u[q], n[d]i(ź)u[q]*, and my favorite, *(m)p[r,l]a[ludzuq]*. And then, how similar to *tik* is *nto'* anyway? These problems of interpretation are legion in the Proto-World work; it is difficult to avoid the conclusion that the data are being shoehorned into a preconceived notion.

Obfuscatory, pedestrian, overcautious quibbles, the Proto-World defenders would object. It is impossible to do science without some frame of reference, some hypothesis one is hoping to see proved. That's true in itself, and in its light, couldn't the sheer weight of tentative protoforms all pointing (*tik*-ing?) in one direction constitute in itself a suggestion of what the Ur-form was? If this protolanguage data really didn't suggest an Ur-reconstruction, then wouldn't we expect there to be *no* visible patterns at all, whether or not the data were complete?

For one thing, we return to the chance point: even taking just a dozen or so family protolanguages, given that each language has several thousand words, we would expect the occasional daffy correspondences to pop up between some of them in some words, even as they were continually turning inside out through the millennia like lava-lamp clumps. But the problem is that even the "patterns" in these protolanguages that Bengtson and Ruhlen point to may only be patterns depending on the eye of the beholder.

Ruhlen claims that a Martian or a layperson looking at the Proto-World data would effortlessly conclude that they indicated forms from the first language, but I'm not so sure. For example, Bengtson and Ruhlen claim that the Proto-World form for "water" is *aq'wa* (the *q* represents a *k* pronounced farther back in the throat). This is based on the following twelve family protolanguage forms (keeping in mind, also, the shaky justification for many of the forms in the first place):

k''ā *nki* *engi* *ak'*^{w}*a* *rt*^{s}*'q'a* *nīru*
ak^{w}*ā* *'oχ*^{w}*a* *namaw* *okho* *gugu* *akwā*

Okay, laypersons and Martians: Do these twelve forms give the immediate appearance of *all* being variations on the same single word to you? *Engi* has a *g* in it, a sound similar to *k*, and vowels on both ends, but then so do hundreds of words in just about any language; the word for "water" in a language could easily fall into that very broad template by accident. *Gugu* could be taken as a kind of perversion of the *kw* sound in *aq'wa* and then repeated to imitate, say, the babbling of a brook–but then so could *klakla* or *koko* or what have you, and if the word were just *ko*, then that, too, could be taken as a truncated *kw*. If *rt*^{s}*'q'a* can be taken as a plausible variation on *aq'wa*, then so could *fakla*, *ik*^{w}*e*, *ingo'*, *praha*, and even our Cheyenne *aa'* (after all, winter has snow, and snow is frozen water . . .)–you get the point. The problem is that so many forms could be taken as "kind of like *aq'wa*" that it is impossible to distinguish between significant and chance resemblances.

Properly, it is safer to say that a layperson or a Martian would see a possible relationship between four of these forms: *ak*^{w}*ā*, *ak'*^{w}*a*, *akwā*, and *'oχ*^{w}*a*. There is no point in dwelling in skepticism for its own sake and trying to argue away these similarities. The problem is that the reasons for these similarities argue not for *aq'wa* as the shape of the world's first word for "water," but for conclusions useful but much less dramatic. The first of those forms is a conventionally accepted Proto-Indo-European reconstruction. The second is presented as the protoform of the Afro-Asiatic family (which includes Arabic, Hebrew, and Nigeria's Hausa); one quibble is that the cited scholar's reconstruction is just *ak'*^{w}*-*; Ruhlen tacks the following *-a* on as a guess based on the fact that some Afro-Asiatic languages have an *-a* in their versions of this word–but then just as many have other vowels. But still, *ak'*^{w}*-* does resemble Proto-Indo-European's *ak*^{w}*ā*. But here, language mixture comes into play: Afro-Asiatic includes Fertile Crescent populations who mingled frequently through the millennia with people ancestral to Proto-Indo-Europeans, and one modern theory even proposes that such peoples all lived and traded around the basin that was later flooded by the Mediterranean and became the Black Sea. But then this resemblance might stem from an ancestor common to Indo-European and Afro-Asiatic. Crucially, however, this merely suggests a Proto-Indo-European/Afro-Asiatic megafamily—quite plausible, given the geographical proximity of the speakers. It is not an argument for Proto-World, leaving the *gugu*'s and *rtˢ'q'a*'s unaccounted for.

The issue is similar with *'oχʷa.* This, as in the Uralic case, is a protoform not for a whole family but for just a sub-*sub*-family of the Caucasian languages spoken in the area best known today as including Chechnya and Georgia, and it means "to drink," not "water." For the Caucasian family as a whole, the closest thing to a "water" protoform is *-ŭGwV* meaning "rain," the *V* symbolizing that we do not know what the vowel was. Still, even if we assume that something like this turned out to be the family protoform for "water," Caucasian is, like Afro-Asiatic, spoken next door to Indo-European, thus meaning that the similarity could be due either to language mixture or to these particular families having an ancestor. In fact, a group of Russian linguists have proposed a "megafamily" called Nostratic, which includes Indo-European, Afro-Asiatic, and one subfamily of Caucasian, among a few others. Again, we remain a long way from Proto-World.

And what happens when the protoform was *not* used by people living near or mingling with Proto-Indo-Europeans? The fourth one, *akwā,* is presented as the protoform of a family including most of the Native American languages of the New World. This one, though, is yet another instance where a form reconstructed for a mere subfamily is presented as grandfather to the family as a whole: horses as Proto-Mammal. *Akwā* is reconstructed only for Cheyenne's Algonquian group, a particularly serious problem in this case because Algonquian is but a drop in the bucket against the dazzling number and variety of other groups of Native American languages, which range from Alaska down to southern South America. Yet Proto-World advocates claim that *akwā* is likely to have been the protoform for "water" for nearly every indigenous language of the Western Hemisphere on the basis of various words scattered among them, typical of which are my hypothetical example on page 297, *koko*, as well as a casserole of other words such as *uk, yok-ha, 'aha', ku'u, iagup, uku-mi, oxi'*, and, well, you get the picture. Moreover, as many of them mean things such as "drink," "go in water," "lake," "he is drinking," and "wet" as actually mean "water." This is no more logically valid than reconstructing what the first cat must have looked like by deducing from pet cats, lions, tigers, etc., and then proposing that this must be what the first mammal looked like–because other mammals have some traits in common with cats. Thus this Algonquian form is much more likely to be a form the word for "water" happened to take in the eternal transformation of this particular sub-sub-sub-sub-subbranch of the first language.

Thus what we see in the list of twelve forms is evidence of a relationship between three families spoken near one another–interesting but hardly Proto-World–and one accident buried within one of dozens of groups of Native American languages.

Then *namaw* and *nīru* are apparently just static, and though one must allow for such things, one must admit that there is a lot of it in the family protolanguage data Bengtson and Ruhlen appeal to. The Proto-World form for "who" is proposed as *ku*, but one family has *min*, another has *yāv*, another has *na*. And then once again, we must recall that the twenty-seven words proposed are not meant to imply that all or even most of the words in the family protolanguages correspond to even this extent. That is, it's not as if, in languages of the Caucasus, words teasingly similar to Indo-European ones like *-ŭGwV* for water are par for the course. Overall, the vocabulary in a Caucasian language such as Georgian is a bizarrely unfamiliar grab bag to any Indo-European speaker. Cross-family correspondences–and, as we have seen, limited, vague, and often questionable ones–are found only in a little bunch of twenty-seven words among the several thousand at least that any one language has. As to whether these likenesses represent an echo of Adam and Eve's vocabulary, we must ask: Why just these words? How would chance allow any particular word to change much, much, much more slowly than the language around it? Or furthermore, what would lead that word to change more slowly not just in one language but in innumerable *separate* ones spread over enormous swatches of the globe, with that particular packet of sounds and its semantic referent somehow retaining a mysterious conservative essence across vastly disparate peoples and languages through countless millennia?

To put a point on it, the notion that Adam and Eve called water *aq'wa* requires us to believe that the Italians, whose word for "water" is *acqua* [AH-kwah], have somehow and for some reason retained this Proto-World form virtually pristine while, inexplicably, all the thousands upon thousands of other words in their offshoot of the first language evolved apace. More precisely, once created in East Africa in 148,000 B.C., *aq'wa* apparently remained inviolate in a particular group among the various offshoot groups that carried the first language northward into the Fertile Crescent, and this one group retained this word frozen solid for tens of thousands of years, generation after generation, *aq'wa* and *aq'wa* alone miraculously impervious to erosion or transformation of sounds while over in Japan people just let the word evolve into *mizu* and while the Cheyenne were turning *peponwi* into *aa'*. The word then survived tumultuous westward migrations into Europe as intact as always. All the other descendants of these Proto-Indo-European-speaking migrants, spreading throughout Europe, started out with this word but immediately began letting it evolve like the rest of their vocabularies, but *aq'wa* was held especially sacrosanct by the peoples who last settled the Italian peninsula and lives on there to this day, as eternal as Orion's Belt. Clearly, one need not be a prosaically minded, cranky drudge brainwashed by "linguistic nonsense" to find this scenario unlikely.

Beyond the Proto-World reconstructions, Joseph Greenberg was one of the seminal figures in twentieth-century linguistics. He brought taxonomic order to the thousand-odd languages of Africa, classifying them into families and subfamilies in a landmark work that has been the foundation for all subsequent work in this area. He also developed a framework for analyzing universal traits and patterns of language structure and development that lies at the heart of the work of hundreds of linguists today. We have seen that some scholars have grouped a handful of families into a "megafamily"; Greenberg, too, made a separate proposal in this vein. Work of this kind is controversial but, in appealing more to correspondences less likely to be due to chance than the Proto-World ones, it is considered plausible and promising by many. Greenberg also proposed a historical taxonomy of Native American languages and tentative "Amerind" protoforms. This work has also met with considerable opposition from traditional historical linguists, but in my view, at the end of the day some of the claims in this work are most likely valid and could not have been uncovered by traditional methods. In fact, the Proto-World idea is just one corner of a career nothing short of monumental overall; it is Ruhlen who has expanded and argued most widely for Proto-World.

Nevertheless, when it comes to the particular question of "how Adam and Eve talked," the dismissal by linguists in media treatments is not just peevish hairsplitting of a sort that blinds people who ought to know better into resisting a magnificent insight sitting right under their noses. Most unfortunately–because I would give quite a bit to know even a handful of words from the first language–linguists' response to Proto-World is based on simple logic. Chaos theoreticians, please work on.

Yet in the end, the reconstruction of Proto-World is not entirely hopeless. Although we will never know its words, through deductive reasoning based on principles we have seen in this book, we can reconstruct certain broad aspects of what the first language's structure must have been like.

Inflections almost always begin as free words; even in the cases where they do not, they arise through sound erosion and change. To wit, inflections always emerge in grammars where they did not exist before. This means that the first language, not having existed for a long enough time for inflections to appear through grammaticalization or other gradual processes, can be assumed not to have had inflections. For the same reason, the first language is unlikely to have been tonal, because tones develop accidentally through change occurring in a grammar where once there was not tone, as we saw with Vietnamese.

We can also reasonably suppose that the first language was low on decorative bells and whistles explicitly marking nooks and crannies of meaning that context would take care of just as well. Alienable possessive marking, definite and indefinite articles, and evidential markers are absent in grammars as often as they are present, and they only come about through reinterpretation of more cognitively fundamental words with the passage of time.

Moreover, the speaker of the first language was probably more likely to say *You want food, go hunt* than *Go hunt if you want food; Animals hunt, eat meat* than *Animals that hunt eat meat*. Because writing had yet to come along to conventionalize them, relative and subordinate clauses were probably not common in speech in the first language, just as they are not in many languages even today, where the preceding ways of putting things would be quite ordinary (Chinese and Vietnamese pattern rather like this, for example).

In other words, of the languages extant today, the ones that most closely approximate the first language are creoles, the only languages that have risen from the ground (pidgins) so recently that they still have not had much time to meander into areas decorative to the needs of human communication. Furthermore, as largely unwritten languages, most creoles have yet to be diverted very far into the particular tics that writing tends to lend a grammar.

More specifically, the first language would have most resembled the creoles that have had the least contact since their birth with the languages that provided their words, such that they have not had occasion to take on aspects typical of older languages from those source languages. This means that about the closest we can get to what that first language was like would be creoles such as Sranan, Tok Pisin, and the Portuguese creoles of the Gulf of Guinea.

This likeness also reveals another likely feature in Proto-World: words would have been easily used as nouns, verbs, adjectives, or adverbs without needing to be "converted" with special affixes. Whereas in English, *heavy* must take *-ness* before serving as a noun (*heaviness*) or be preceded by *make* to serve as a verb (*to make heavy*), in Sranan, *heaviness*, *make heavy*, and *heavy* itself are all *hebi*. *A saka hebi! A hebi e-hebi mi!* would be a perfectly legal way of saying *The bag is heavy! The weight is weighing me down!* This is particularly common in creoles because they have not existed long enough to give rise to inflections and other constructions that the grammar can easily do without, which is how such conversion strategies appear in languages such as English. It is almost certain that words in the first language, which was similarly new to the world, had a flexibility of word-class usage of the kind common in creoles.

Of course, as I noted before, even creoles have their "bells and whistles," such as frequently having articles, as well as making extensive use of tense markers, which all languages do not and which suggests that even this seemingly sine qua non aspect of speech is actually an add-on as well. However, taking the general outline of creole structure and then deducing even further in reverse to reconstruct what a grammar would look like the second the lava lamp was turned on, we can glean a sense of what the first language's grammar was like.

The first language most likely had no inflections or tones. It overtly marked only what is necessary to communicating, without delving into such features as whether something has already been mentioned (that is, articles), shades of intensity of possession, or just how one heard something. In line with this, tense was probably marked only when explicitly necessary, with context taking care of the rest. The first language's way of packaging information would have been relatively "telegraphic" in comparison with European languages. Finally, a word could be used in a different word class (noun as a verb, verb as an adjective, etc.) without needing to undergo any changes in the process.

This means that ultimately our closest living approximation to human language as a lily ungilded is not words like *tik* and *aq'wa* but the grammars of certain creole languages largely unknown beyond where they are spoken. In the languages that today get the most press–a dozen or so in Europe, plus Chinese, Japanese, and Arabic–the structure of the first language remains only as shadows and vestiges, the rest long since obscured by morphings and random accretions upon the foundational rootstuff. The living languages most like the parent of all six thousand are spoken in places as little known to most of the world as Surinam, islands off the west coast of Africa, and Papua New Guinea.

Notes

Introduction

5 Qualitative difference between human and animal speech: Derek Bickerton, *Language and Human Behavior* (Seattle: University of Washington Press, 1995), chap. 1; and Terrence W. Deacon, *The Symbolic Species* (New York: Norton, 1997), pp. 28–32.

7 Chinese hominids, evidence for language origin: see Bickerton, *Language and Human Behavior,* p. 70.

8 Evidence for innateness of language: Steven Pinker, *The Language Instinct* (New York: Morrow, 1994), esp. chap. 10.

8 Evidence against: see Deacon, *Symbolic Species;* and Geoffrey Sampson, *Educating Eve* (London: Cassell, 1997).

1 The First Language Morphs into Six Thousand New Ones

17 Blowing up the post office: Patricia Cline Cohen, *The Murder of Helen Jewett* (New York: Vintage, 1998), p. 141.

24 Nabokov's *terrible:* Vladimir Nabokov, "Father's Butterflies," trans. Dmitri Nabokov, *The Atlantic Monthly,* April 2000, p. 75.

39 *Hamlet* translations:

French: *La tragique histoire d'Hamlet,* trans. André Gide, in *Oeuvres complètes de Shakespeare,* vol. 2 (Tours: Gallimard, 1959).

Spanish: *Hamlet* in Guillermo Shakespeare, *Hamlet / Romeo y Julieta,* trans. Leandro Moratín and M. Menéndez y Pelayo (Madrid: Biblioteca E.D.A.F., 1964).

Italian: *Amleto,* trans. Raffaello Piccoli, in *William Shakespeare: Teatro,* vol. 2 (Florence: Sansoni, 1957).

German: *Hamlet, Prinz von Dänemark: Ein Trauerspiel,* trans. Christoph Martin Wieland (Zürich: Haffmanns, 1995).

Russian: *Gamlet: Prints Damskij,* trans. M. Lozinskovo (Moscow: Detskaja Literatura, 1965).

Hebrew: *Hamlet,* trans. Avraham Shlonsky (Merhavyah: Sifriyat Poalim, 1971).

40 "Fratin" comment: Hubert Monteilhet, *Neropolis: Roman des temps néroniens* (Paris: Éditions du Juillard, 1984), pp. 159–160.

45 *Sister-in-law* in Indo-European: Derived from Calvert Watkins, "Indo-European and the Indo-Europeans," an appendix in *The American Heritage Dictionary of the English Language,* 3rd ed. (Boston: Houghton Mifflin, 1992), pp. 2081–2082.

50 Javanese varieties: Clifford Geertz, "Linguistic Etiquette," in *Sociolinguistics,* ed. John Pride and Janet Holmes (Harmondsworth, England: Penguin, 1972), pp. 167–179.

2 The Six Thousand Languages Develop into Clusters of Sublanguages

56 Cornwall English: Ian Hancock, "Componentiality and the Creole Matrix," in *The Crucible of Carolina,* ed. Michael Montgomery (Athens: University of Georgia Press, 1994), pp. 95–114.

64, 65 Quotations on the triumph of Standard French: Ralph D. Grillo, *Dominant Languages* (Cambridge: Cambridge University Press, 1989), pp. 31–35.

66, 67 Caxton quotation and quotation from play (from *The Towneley Cycle*): David Crystal, *The Cambridge Encyclopedia of the English Language* (Cambridge: Cambridge University Press, 1995), pp. 55, 57.

67 Dante quotations: Daniel Boorstin, *The Creators* (New York: Vintage, 1992), pp. 255–264.

68 On Moldovan: Donald L. Dyer, *The Romanian Dialect of Moldova* (Lewiston, NY: Mellen Press, 1999.)

72 Literary excerpt on Ukrainian: Edward Rutherfurd, *Russka* (New York: Ivy Books, 1991 [paperback edition]), p. 639.

74 Wuhan Chinese: Genevieve Escure, *Creole and Dialect Continua* (Amsterdam: Benjamins, 1997), p. 146.

76 *Asterix* scenes:

German version: Albert Uderzo, *Der Grosse Graben,* trans. Gudrun Penndorf from French original, *Le grand fossé* (Stuttgart: Egmont Ehapa, 1980).

Schwäbisch version: Albert Uderzo, *Dr grosse Graba,* trans. Klaus-Dieter Mühlsteffen (Stuttgart: Egmont Ehapa, 1998).

Swiss German version: Albert Uderzo, *Dr gross Grabe,* trans. Hansruedi Lerch (Stuttgart: Ehapa, 1996).

77 New Britain "languages": William R. Thurston, *Processes of Change in the Languages of North-Western New Britain* (Canberra: Pacific Linguistics B99, 1987), pp. 28–29.

78–79 Scots passages: Prodigal Son passage by William Laughton Lorimer and excerpt from "The Actes and Life of the most Victorious Conqueror, Robert Bruce King of Scotland" by John Barbour in Crystal, *Cambridge Encyclopedia of the English Language* (p. 52).

80–81 Gurage varieties: Wolf Leslau, *Etymological Dictionary of Gurage (Ethiopic)* (Wiesbaden: O. Harrassowitz, 1979).

81 African language continuum: Adolphe Dzokanga, *Dictionnaire Lingala-français suivi d'une grammaire Lingala* (Leipzig: VEB Verlag Enzyklopädie, 1979), p. 6.

85 Evenki and Oruqen: Lindsay J. Whaley, Lenore A. Grenoble, and Fengxiang Li, "Revisiting Tungusic Classification from the Bottom Up: A Comparison of Evenki and Oruqen," *Language* 75 (1999): 286–321.

86–87 Old English dialects: see Crystal, *Cambridge Encyclopedia of the English Language,* p. 29.

89 Bavarian *Asterix:* René Goscinny and Albert Uderzo, *Auf geht's zu de Gotn!* trans. Well Hans (Stuttgart: Delta, 1997).

3 The Thousands of Dialects Mix with One Another

98 Russians learning English: Boris Startzev, "Štob Epigrafy Razbirat'," *Itogi,* January 25, 2000, pp. 53–54.

101 Peter the Great: Viktor V. Vinogradov, *Istoria Russkovo Literaturnovo Jazjka* (Moscow: Nauka, 1978), p. 44.

102 Fifty percent vocabulary borrowing in Australia: Robert M. W. Dixon, *The Rise and Fall of Languages* (Cambridge: Cambridge University Press, 1997), pp. 26–27.

103 Norman versus Parisian French in English: David Crystal, *The Cambridge Encyclopedia of the English Language* (Cambridge: Cambridge University Press, 1995) p. 46.

104 French in Scots: see Crystal, *Cambridge Encyclopedia,* p. 52.

104 Scots Prodigal Son: see Crystal, *Cambridge Encyclopedia,* p. 328.

107 Balkan Sprachbund: Hans Heinrich Hock, *Principles of Historical Linguistics* (Berlin: Mouton de Gruyter, 1991), pp. 494–498.

107 Kupwar Sprachbund: John J. Gumperz and Robert Wilson, "Convergence and Creolization: A Case from the Indo-Aryan / Dravidian Border in India," in *Pidginization and Creolization of Languages,* ed. Dell Hymes (Cambridge: Cambridge University Press, 1971), pp. 151–167.

112 Michif: The most useful single source is Peter Bakker, *A Language of Our Own* (New York: Oxford University Press, 1997).

113 Media Lengua: Pieter Muysken, "Media Lengua," in *Contact Languages: A Wider Perspective,* ed. Sarah G. Thomason (Amsterdam: Benjamins, 1997), pp. 365–426.

114 Mednyj Aleut: Evgenij V. Golovko, "Copper Island Aleut," in *Mixed Languages,* ed. Peter Bakker and Maarten Mous (Amsterdam: IFOTT, 1994), pp. 113–121.

114 Gypsy intertwined languages: Norbert Boretzky and Birgit Igla, "Romani Mixed Dialects," in Bakker and Mous, eds., *Mixed Languages,* pp. 35–68.

115 Isicamtho: G. Tucker Childs, "The Status of Isicamtho, an Nguni-Based Urban Variety of Soweto," in *The Structure and Status of Pidgins and Creoles,* ed. Arthur K. Spears and Donald Winford (Amsterdam: Benjamins, 1997), pp. 341–470.

115 Wutun: Mei W. Lee-Smith and Stephen A. Wurm, "The Wutun Language," in *Atlas of Languages of Intercultural Communication in the Pacific, Asia, and the Americas,* ed. Stephen A. Wurm, Peter Mühlhäusler, and Darrell T. Tyron (Berlin: Mouton de Gruyter, 1996), pp. 883–897.

116 Greek in Turkey: Sarah Grey Thomason and Terrence Kaufman, *Language Contact, Creolization, and Genetic Linguistics* (Berkeley: University of California Press, 1988), pp. 215–222.

117 English incursions into German: Jürgen Steinhoff, "Sprachstörung," *Stern,* September 2, 1999.

120 Spanglish and the Internet: Sam Dillon, "On the Language of Cervantes, the Imprint of the Internet," *New York Times,* August 6, 2000, Week in Review, p. 3.

120 Immigrant Russian: Alexei Yurchak, "Protsess Bystrovo Izmenenija Jazyka i Kul'tury," *Kabinet* (October 1995).

121 John Cheke on borrowings in English: see Crystal, *Cambridge Encyclopedia of the English Language,* p. 125.

122 Language "stewing" as default: Robert M. W. Dixon, *The Rise and Fall of Languages* (Cambridge: Cambridge University Press, 1997), pp. 68–96.

125 Europe as a Sprachbund: Martin Haspelmath, "How Young Is Standard Average European?" *Language Sciences* 20 (1998): 271–287.

128, 129 DNA transfer, macramé analogy: Richard Lewontin, *It Ain't Necessarily So: The Dream of the Human Genome and Other Illusions* (New York: New York Review of Books, 2000), pp. 71–72.

128 Margulis quotation: Lynn Margulis, *Symbiotic Planet: A New View of Evolution* (New York: Basic Books, 1998), p. 52.

4 Some Languages Are Crushed to Powder but Rise Again as New Ones

There are two books that are invaluable sources for accessibly written information on pidgins and creoles: John Holm's *Pidgins and Creoles,* vol. 2 (Cambridge: Cambridge University Press, 1989), an encyclopedia of all pidgins and creoles substantially documented by the time of its publication; and Mark Sebba's *Pidgins and Creoles* (New York: St. Martin's Press, 1997), a textbook aimed at the general reader that covers the territory so perfectly that I canceled plans to write a textbook of my own on pidgins and creoles.

134 Russenorsk: Ingvild Broch and Ernst Håkon Jahr, "Russenorsk: A New Look at the Russo-Norwegian Pidgin in Northern Norway," in *Scandinavian Language Contacts,* ed. P. Sture Ureland and Iain Clarkson (Cambridge: Cambridge University Press, 1984), pp. 21–65.

136 Native American Pidgin English passage: Douglas Leechman and Robert A. Hall, Jr., "American Indian Pidgin English: Attestations and Grammatical Peculiarities," in *Perspectives on American English,* ed. J. L. Dillard (The Hague: Mouton de Gruyter, 1980), p. 419.

136 *Tintin* sequence: Hergé, *Tintin au Congo* (Paris: Casterman, 1946), p. 20.

138 South Seas Pidgin English sentences: Roger Keesing, *Melanesian Pidgin and the Oceanic Substrate* (Palo Alto, CA: Stanford University Press, 1988), pp. 14, 31–32.

140–141 Tok Pisin passage: Mark Sebba, *Contact Languages* (New York: St. Martin's Press, 1997), pp. 20–21.

142 Tok Pisin *bai* passage: Gillian Sankoff and Suzanne Laberge, "On the Acquisition of Native Speakers by a Language," in *The Social Life of Language,* ed. Gillian Sankoff (New York: Academic Press, 1980), p. 204.

142–143 Tok Pisin's five tense markers: Peter Mühlhäusler, *Pidgin and Creole Linguistics,* expanded rev. ed. (London: University of Westminster Press, 1997), p. 171.

144 Tok Pisin "make it change" marking: see Mühlhäusler, *Pidgin and Creole Linguistics,* p. 170.

144 Tok Pisin word-meaning nuances: see Mühlhäusler, *Pidgin and Creole Linguistics,* pp. 183, 185.

145 Tok Pisin spoken natively by children: see Sankoff and Laberge, "Acquisition of Native Speakers," p. 199.

148 Annobonese creole Portuguese sentence and function of *xa* and *sa:* Marike Post, "Aspect Marking in Fa d'Ambu," in *From Contact to Creole and Beyond,* ed. Philip Baker (London: University of Westminster Press, 1995), pp. 195–196.

149 Louisiana French creole sentence: Albert Valdman and Thomas A. Klingler, "The Structure of Louisiana Creole," in *French and Creole in Louisiana* (New York: Plenum, 1997), p. 111.

150 Chinook creole: Anthony P. Grant, "The Evolution of Functional Categories in Grand Ronde Chinook Jargon: Ethnolinguistic and Grammatical Considerations," in *Changing Meanings, Changing Functions,* ed. Philip Baker and Anand Syea (London: University of Westminster Press, 1996), pp. 234–235.

152 Solomon Islands Pijin–Kwaio comparison: see Keesing, *Melanesian Pidgin,* p. 121.

159 Berbice Dutch sentence: John Holm, *Pidgins and Creoles,* vol. 2 (Cambridge: Cambridge University Press, 1989), p. 333.

162 Réunion French sentence: Robert Chaudenson, *Le lexique du parler créole de la Réunion* (Paris: Champion, 1974), p. 1165.

164 Guyanese creole continuum: William R. O'Donnell and Loreto Todd, *Variety in Contemporary English* (London: Allen & Unwin, 1980), p. 52.

168 Fula data: William E. Welmers, *African Language Structures* (Berkeley: University of California Press, 1973), p. 203.

168 Nonnative Fula: Pierre-François LaCroix, "Quelques aspects de la désintégration d'un système classificatoire (peul du sud de l'Adamawa)," in *La classification nominale dans les langues négro-africaines,* Colloques internationaux du Centre National de la Recherche Scientifique (Paris: Éditions du Centre National de la Recherche Scientifique, 1967), pp. 291–308.

170 Philippines creole Spanish sentence courtesy of John Wolff and Margaret Ong.

172 Fly Taal: see Holm, *Pidgins and Creoles,* pp. 350–352.

5 The Thousands of Dialects of Thousands of Languages All Develop Far Beyond the Call of Duty

180 Tuyuca evidentials: Janet Barnes, "Classifiers in Tuyuca," in *Amazonian Linguistics: Studies in Lowland South American Languages,* ed. Doris Payne (Austin: University of Texas Press, 1990), pp. 273–292.

181 Makah evidentials: William H. Jacobsen, "The Heterogeneity of Evidentials in Makah," in *Evidentiality: The Linguistic Coding of Epistemology,* ed. Wallace Chafe and Johanna Nichols (Norwood, NJ: Ablex, 1986), pp. 3–28.

182 Navajo possession: Robert W. Young and William Morgan, *The Navajo Language: A Grammar and Colloquial Dictionary* (Albuquerque: University of New Mexico Press, 1980), p. 7.

184 Estimate of cross-linguistic distribution of articles: Edith Moravcsik, "Determination," *Working Papers on Language Universals* 1 (1969): 64–130.

187 Cantonese classifiers: Stephen Matthews and Virginia Yip, *Cantonese: A Comprehensive Grammar* (London: Routledge, 1994), p. 103.

188 Fula grammar: D. W. Arnott, *The Nominal and Verbal Systems of Fula* (London: Oxford University Press, 1970), pp. 67–130.

190 Lewontin quotations: Richard Lewontin, *It Ain't Necessarily So: The Dream of the Human Genome and Other Illusions* (New York: New York Review of Books, 2000), pp. 57–58.

193 Ending loss and adverb placement: Ian Roberts, *Verbs and Diachronic Syntax* (Dordrecht: Kluwer, 1992).

194 Cantonese tones: see Matthews and Yip, *Cantonese,* p. 21.

195 Emergence of tone in Vietnamese: adapted (quite creatively) from André-Georges Haudricourt, "De l'origine des tons en viêtnamien," *Journal Asiatique* 242 (1954): 68–82.

197 *No Small Affair:* Don't rent it.

198 Maori possession: Winifred Bauer, *Maori* (London: Routledge, 1993), pp. 209–213.

200 Cree sentence: Thomas Payne, *Describing Morphosyntax* (Cambridge: Cambridge University Press, 1997) p. 212.

202 Welsh mutations: Martin Ball and Nicole J. Müller, *Mutation in Welsh* (London: Routledge, 1992), pp. 195–196; and Gareth

King, *Modern Welsh: A Comparative Grammar* (London: Routledge, 1993), p. 81.

207 *Asterix:* Albert Uderzo, *Asterix und Maestria,* German translation of *La Rose et le Glaive* (Stuttgart: Ehapa, 1991).

208 Tok Pisin *-pasin* uses: Peter Mühlhäusler, "The Scientific Study of Tok Pisin: Language Planning and the Tok Pisin Lexicon," in *Handbook of Tok Pisin (New Guinea Pidgin),* ed. Stephen A. Wurm and Peter Mühlhäusler (Canberra: Australian National University Press, 1985), p. 625.

209 Structural definition of creoles: John H. McWhorter, "Defining Creole as a Synchronic Term," in *Degrees of Restructuring in Creole Languages,* ed. Ingrid Neumann-Holzschuh and Edgar Schneider (Amsterdam: Benjamins, 2000), pp. 85–123.

212 Development of Hawaiian "Pidgin": Derek Bickerton, *Roots of Language* (Ann Arbor, MI: Karoma, 1981).

6 Some Languages Get Genetically Altered and Frozen

217 Ngan'gityemerri: Nicholas Reid, "Phrasal Verb to Synthetic Verb: Recorded Morphosyntactic Change in Ngan'gityemerri," in *Studies in Comparative Non-Pama Nyungan,* ed. Nicholas Reid (Canberra: Pacific Linguistics, 1999).

230 *You was* letters: Patricia Cline Cohen, *The Murder of Helen Jewett* (New York: Vintage, 1998), pp. 242–245.

232 History of Standard Russian: Viktor V. Vinogradov, *Istoria Russkovo Literaturnovo Jazyka* (Moscow: Nauka, 1978); and G. O. Vinokur, *The Russian Language: A Brief History* (Cambridge: Cambridge University Press, 1971).

233 "One converses in Russian . . .": see Vinogradov, *Istoria Russkovo Literaturnovo Jazyka,* p. 39.

233 Development in Russian of *e* to *o:* see Vinokur, *Russian Language,* pp. 69–70.

237 Baldwin passage: James Campbell, *Talking at the Gates* (New York: Viking, 1991), p. 143.

240 Written versus spoken language: a useful summary of this issue is Wallace Chafe's "Linguistic Differences Produced by Differences Between Speaking and Writing," in *Literacy, Language, and Learning,* ed. David R. Olson, Nancy Torrance, and Angela Hildyard (Cambridge: Cambridge University Press, 1985), pp. 105–123.

240 Malory passage: from Chapter 8, Book XIII, of *Morte d'Arthur,* cited by David Crystal, *The Cambridge Encyclopedia of the English Language* (Cambridge: Cambridge University Press, 1995), p. 58.

242 Old Russian passage: see Vinokur, *Russian Language,* p. 66.

243 Modern Russian passage: Mikhail Bulgakov, *Master i Margarita* (1940; reprint, Moscow: Olimp, 1996), p. 30. Translation: *The Master and Margarita,* trans. Michael Glenny (London: Harvill, 1967), p. 27.

244 Somali: Douglas Biber and Mohamed Hared, "Linguistic Correlates of the Transition to Literacy in Somali: Language Adaptation in Six Press Registers," in *Sociolinguistic Perspectives on Register,* ed. Douglas Biber and Edward Finegan (New York: Oxford University Press, 1994), pp. 182–216.

247 Inuktitut: Ivan Kalmár, "Are There Really No Primitive Languages?" in *Literacy, Language and Learning,* ed. David R. Olson, Nancy Torrance, and Angela Hildyard (Cambridge: Cambridge University Press, 1985), pp. 148–166.

248 No subordinate clauses in Proto-Indo-European: Eduard Hermann, "Gab es im Indogermanischen Nebensätze?" *Zeitschrift für vergleichende Sprachforschung* 33 (1895): 481–535.

248 Tirió ritual language: Mark J. Plotkin, *Tales of a Shaman's Apprentice* (New York: Viking, 1993), p. 104.

249 Jamaican ritual language: Kenneth Bilby, "How the 'Older Heads' Talk: A Jamaican Maroon Spirit Possession Language and Its Relationship to the Creoles of Suriname and Sierra Leone," *Nieuwe West-Indische Gids* 57 (1983): 37–88.

250 Underwater views: Stephen Jay Gould, "Seeing Eye to Eye, Through a Glass Clearly," in *Leonardo's Mountain of Clams and the Diet of Worms* (New York: Three Rivers Press, 1998).

250 Prototypical "written English" passage: Chafe, "Linguistic Differences," p. 107.

7 Most of the World's Languages Went Extinct

254 Eleven o'clock: Daniel Nettle and Suzanne Romaine, *Vanishing Voices: The Extinction of the World's Languages* (New York: Oxford University Press, 2000), pp. 103–104.

256 Maa: Gerrit J. Dimmendaal, "On Language Death in Eastern Africa," in *Investigating Obsolescence: Studies in Language Contraction and Death,* ed. Nancy Dorian (Cambridge: Cambridge University Press, 1989), pp. 20–23.

257 Top twenty languages: David Crystal, *The Cambridge Encyclopedia of Language* (Cambridge: Cambridge University Press, 1987), p. 287.

257 Ninety-six percent of the world speaks the top twenty languages: David Crystal, *Language Death* (Cambridge: Cambridge University Press, 2000), p. 14.

257 Ninety percent of the world's languages will die: Michael Krauss, "The World's Languages in Crisis," *Language* 68 (1992): 4–10.

259 "Linguistic equilibrium": proposed first by Robert M. W. Dixon, *The Rise and Fall of Languages* (Cambridge: Cambridge University Press, 1997), pp. 68–73.

259 The agricultural revolution and its effect on language spread: Jared Diamond, *Guns, Germs, and Steel* (New York: Norton, 1997).

262 Cayuga decay: Marianne Mithun, "The Incipient Obsolescence of Polysynthesis: Cayuga in Ontario and Oklahoma," in *Investigating Obsolescence: Studies in Language Contraction and Death,* ed. Nancy Dorian (Cambridge: Cambridge University Press, 1989), pp. 247–248.

263–264 Pipil inflections: Lyle Campbell and Martha C. Muntzel, "The Structural Consequences of Language Death," in *Investigating Obsolescence,* ed. Dorian, pp. 192–193.

264 Scottish Gaelic: Nancy C. Dorian, "Grammatical Change in a Dying Dialect," *Language* 49 (1973): 413–438.

266 Cayuga verb–noun complexes: see Mithun, "Incipient Obsolescence of Polysynthesis," pp. 249–250.

267 Breton dialects and the revival movement: Lois Kuter, "Breton vs. French: Language and the Opposition of Political, Economic, Social, and Cultural Values," in *Investigating Obsolescence,* ed. Dorian, pp. 84–85.

268 Gros Ventre: Allan R. Taylor, "Problems in Obsolescence Research: The Gros Ventres of Montana," in *Investigating Obsolescence,* ed. Dorian, p. 176.

269 Mohawk sentence: Marianne Mithun, "The Significance of Diversity in Language Endangerment and Preservation," *Endangered Languages: Language Loss and Community Response,* ed. Lenore A. Grenoble and Lindsay J. Whaley (Cambridge: Cambridge University Press, 1998), p. 177.

271 "Hayseed dialects": Frederick Bodmer, *The Loom of Language* (New York: Norton, 1944), p. 346.

275 Ladefoged and Dorian articles: Peter Ladefoged, "Another View of Endangered Languages," *Language* 68 (1992): 809–811; Nancy C. Dorian, "A Response to Ladefoged's Other View of Endangered Languages," *Language* 69 (1993): 575–579.

275 Nettle and Romaine book: see Nettle and Romaine, *Vanishing Voices.*

Epilogue

287 Sources on Proto-World: Merritt Ruhlen, *The Origin of Language* (New York: Wiley, 1994); and John D. Bengtson and Merritt Ruhlen, "Global Etymologies," in *On the Origin of Languages: Studies in Linguistic Taxonomy,* ed. Merritt Ruhlen (Stanford, CA: Stanford University Press, 1994).

288 Ruhlen quotation: Ruhlen, *Origin of Language*, p. 133.

294 Paper showing statistical counterevidence to Proto-World argumentation: William Poser, "The Mathematics of Multilateral Comparison," presented at the Annual Meeting of the Linguistic Society of America, New Orleans, January 5, 1995.

295 Uralic and Miao-Yao regarding Proto-World: Joseph Salmons, "A Look at the Data for a Global Etymology: **tik* 'finger'," in *Explanation in Historical Linguistics,* ed. Garry W. Davis and Gregory K. Iverson (Amsterdam: Benjamins, 1992), pp. 206–228.

Acknowledgments

I am eternally thankful to my friends in linguistics and beyond who are always so helpful with my out-of-the-blue requests for obscure information and advice and so diligent in keeping my data honest: in this case, Ronelle Alexander, Michelle Bazu, Hans Boas, Sue Wen Chiao, Donald Dyer, Benny Hary, Leanne Hinton, Gary Holland, Darya Kavitskaya, Marianne Mithun, John Ohala, Irina Paperno, Mikael Parkvall, Maria Polinsky, Richard Rhodes, Paula Rogers, Hans-Jürgen Sasse, and Engin Sezer.

I wish I could claim that I came up with the title of the book, but in fact it was Curtis Bartosik who did so at a party, taking a quick walk around the block in the chill of a Bay Area evening and coming back with "The Power of Babel" seemingly picked right out of the air.

Special thanks to Ashlee Bailey, for formatting the charts in Chapter 5 and for feedback on early chapter drafts, and to my fellow bibliomaniac and "language-head" Stéphane Goyette, for reading first drafts of some of the chapters and saving me from pontificating about why French is weird.

My editor, Erika Goldman, shepherded this book into fruition with her accustomed diligence and friendship; and Project Editor Mary Louise Byrd's sharp eyes lent the book a polish that will enable me to open it in ten years without needing to first fortify myself with a drink.

Meanwhile, any "dings" or overgrowths in this book are, of course, strictly *culpa mea*.

Index

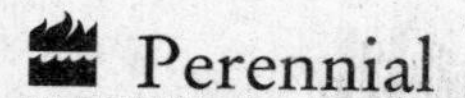

Books by John McWhorter:

LOSING THE RACE

Self-Sabotage in Black America
ISBN 0-06-093593-6 (paperback)

A young African-American scholar offers an explosive diagnosis of the reason blacks lag behind in performance and achievement: contemporary black identity is so rooted in defeatism that blacks have become their own worst enemies in the struggle for success.

"A sincere call to face the unpleasant truths behind black underachievement. If these debilitating trends are ever to be reversed, [McWhorter's] call must be heeded."

—*The Wall Street Journal*

THE POWER OF BABEL

A Natural History of Language
ISBN 0-06-052085-X (paperback)

The first book written for the layperson about the natural history of language. Linguistic professor John McWhorter draws from linguistic theory, geography, history, and pop culture to tell the fascinating story of how thousands of very different languages have evolved from a single, original source in a natural process similar to biological evolution.

"McWhorter explains clearly how and why sounds change, how word meanings change...how grammar changes and how they all bifurcate, mix, multiply, grow branches, get elaborated, are dissolved and reconstituted. McWhorter writes lucidly; it's evident that he's a teacher." —*San Francisco Chronicle*

Available wherever books are sold, or call 1-800-331-3761 to order.